The Memory Project

by

Gerald T. Rainey

TELEMACHUS PRESS

This book is a work of fiction. Names, characters, places and incidents are either the product of the author's imagination or are used fictitiously. Any resemblance to actual persons, living or dead, or to actual events or locales is entirely coincidental.

THE MEMORY PROJECT

The publisher does not have any control over and does not assume any responsibility for author or third-party websites or their content.

Cover design and front cover image by Joneile Emery

Back cover art:
Copyright © Shutterstock/101527918/pr2is

Published by Telemachus Press, LLC
http://www.telemachuspress.com

Visit the author website:
http://www.geraldrainey.com

ISBN: 978-1-938701-57-3 (eBook)
ISBN: 978-1-938701-58-0 (Paperback)

Version: 2014.03.11

Printed in the United States of America

10 9 8 7 6 5 4 3 2 1

"I will never leave you or abandon you ..."
Hebrews 13:5

Table of Contents

THE MEMORY PROJECT

Introduction

THE WEATHER WAS miserable to some; perfect to others. It was raining steadily. A low front was swirling counterclockwise off the western coast of Canada pulling the warm, moist tropical air in from Hawaii. It was a wet storm providing ample moisture to soak the Pacific Northwest and Seattle, carrying on to the interior of the state. The small town of Quincy was in for a drenching.

Downtown Quincy was eerily quiet. It was a ghost town. The cars that normally drove by the city park were gone. The playgrounds sat empty. There were no dogs roaming the streets. The only sound one could hear was the clicking of the street lights as they changed colors for no cars, and the pounding rain.

Highway 28 was closed at Palisades Road SW and Highway 283. Highway 281 was closed at Interstate 90. The remaining country roads were closed a minimum of ten miles from the city. Military personnel were stationed at each roadblock, fully armed, with two tanks. No one was allowed past the roadblocks. No one.

From the point of each roadblock, two soldiers were stationed one hundred yards apart in a vehicle or ATV encircling the entire town. A Humvee with a mounted 50 caliber machine gun was stationed each mile. The combination created an impenetrable defense around the small town where no one could enter or leave from any direction.

A gray, unobtrusive modular building sat at the corner of a field about three miles south of town. It had no windows and two doors. It was positioned next to a power pole near the side of a country road. Military vehicles were scattered throughout the field. Soldiers were placed as sentries at various points around the building.

A Humvee command vehicle with two flags flying on it pulled up to the curb in front of the building. Major General James Ritchie jumped out of the vehicle and walked briskly up to the door followed by his assistant, Lieutenant Rigby. The General was wearing camouflage fatigues with two stars on each collar, the 101st Airborne insignia on his shoulder, shiny boots and a cap. He was carrying a satchel under his arm. He was greeted by several soldiers and a news crew.

Major General Ritchie pointed to the news crew. "Get them out of here."

A newswoman with a microphone held it out to the general. "General, can you tell us what …"

With one swift move, the General grabbed the microphone out of the woman's hand, unplugged it and put it in his pocket. The woman was aghast. "This is a military operation, ma'am," General Ritchie said as he smiled to the woman and continued. "You need to leave now … unless you want me to take your camera too."

The young lady spun on her heels as her cameraman lowered his camera and left.

The General reached into his pocket and handed the microphone to one of the soldiers. "Here, I don't need this." He walked past the soldier and into the building followed by his aide. The soldier stood looking at the microphone. He looked at the other soldier, glanced around and tossed it into the bushes.

The mobile command of the 101st Airborne had arrived.

Someone in the room yelled, "Atten-hut!"

"At ease, people." The General walked over to the head of the table and tossed his satchel onto it. He stood there looking at the map to his right. His aide sat in the chair next to him and opened the satchel. "Where are we with this operation, Sergeant?"

"Sir." The Sergeant was in his mid-forties, built like a rock, and stood tall. Several tours of Iraq and Afghanistan wore like a story across his weathered face. "The town has been evacuated." The Sergeant glanced down at his notes lying on the table. "There has been some resistance, but our troops were able to extract the people without casualties. There have been no reports of significant injuries, sir."

"Fine." The General glanced through some notes. "Grid search?"

"Completed. The target area has been cleared."

"Weather cooperating?"

A Major in the room spoke. "Yes, sir. Heavy rains expected all day into Thursday. We will have effective rainout, sir, reducing the contamination perimeter significantly."

"Outlying areas?"

"There is a mandatory evacuation downwind for ten miles," the Major said. "Even though we expect the rainout to contain the fallout, our troops have informed me that all residents have been relocated to temporary facilities near Moses Lake until the area is safe to re-enter."

"Perimeter security?" the General continued.

The Captain next to the Sergeant spoke. "Secure, General. We have blocked off all roads leading to the target area. Troops have been stationed every hundred yards encircling the impact zone." He walked over to the map posted on the whiteboard. He pointed to the concentric circles centered just to the northeast of Quincy. "For the twenty mile circumference, we have 350 stations." The Captain circled the area with his finger. "This is the entire blast zone. A 340 kiloton bomb will exert fifteen PSI[1] for just over a mile from this point." The Captain pointed to the center of the circles. "From there we have five PSI at two miles, one PSI at five miles." The Captain paused. "Quincy is in the two PSI range, General."

"I see, Captain." The General studied the map. "Do we have confirmation that the drop zone is maximum coverage of the affected area?"

The Captain responded. "We do, General."

---

1 PSI—Pounds per Square Inch

One of the soldiers turned around to face the men standing around the table and the map. He was holding a phone receiver in his hand. "General. It's Lieutenant General Mire from the Pentagon, sir."

"Put him on speaker."

"Jim. Doug, here." The voice over the phone was clear and forceful.

"You're coming in loud and clear. I have you on speaker in OPS."

"Fine. Operation Clean Sweep is a go." Everyone in the room listened intently.

"Roger. Operation Clean Sweep is a go for 1200 hours," the General repeated.

"God's speed." The line went dead.

The men in the room looked at the General, waiting for the order.

"Gentlemen. I wanted all of you to hear the green light. We are about to detonate a nuclear bomb near a city in the United States of America." The General stood to his feet. All eyes were on him. No one moved. "There is a significant threat to human life in this area. I don't fully understand the specifics of what it is. I don't need to. What I need to know is that the President has directed us to take this action. He, and his advisors, believe this is the best, most efficient, most appropriate way to deal with this threat. In one hour, we will be dropping a 340 kiloton B61 low level nuclear bomb three miles northeast of Quincy." The General pointed to the center of the concentric circles on the map. He turned and slowly looked at each soldier in the room, gauging their expressions to see if anyone would falter. The men were steadfast, silent, expressionless.

The General continued. "Fallout contamination in the area is significantly reduced because of the weather. However, there is a high probability the town will be rendered useless. If the force of the blast doesn't destroy the buildings, the SREMP[2] will likely fry everything in the area." The General paused to make sure everyone in the room understood their mission. "We have not been charged to protect the town, but to protect the people … and a nation. Sergeant!"

"Yes, sir!"

---

2 SREMP—Source Region Electro-Magnetic Pulse ,which is produced by low-altitude nuclear bursts.

"Pull your remaining troops out."

"Yes, sir!"

"And get those reporters out of here." The General looked at the map. "We have a mission to complete."

—*—

The pilot was strapped into the F16 Falcon cruising at mach one. He was headed west toward Quincy. The clouds below were full of moisture and riding low in the air currents. The ceiling was just 2,500 feet. The pilot would have but a few seconds to release his payload and speed out of range. This was the first time he ever dropped a B61 nuclear bomb. It was also the first time he ever bombed a city in his own country.

"Red Viper to command."

"Go Red Viper." The General listened to the radio transmissions to the tower and his command center simultaneously. Everyone in the room was listening. He was watching a live broadcast from the nose of the plane. All he could see were clouds below the plane and blue sky above.

"I am one minute to target. Repeat, one minute to target."

"Roger, Red Viper. One minute to target," came the reply.

The men in the room looked at the General. The General looked at the map and a radio operator to his right. The operator shook his head 'no.' The General took the mouthpiece to the radio. "Roger that, Red Viper. This is General Ritchie. Alpha alpha zulu tango niner. Clean Sweep is still a go."

"Affirmative, sir. Alpha alpha zulu tango niner confirmed. Clean Sweep is a go," The pilot repeated.

The General turned away from the mouthpiece. "Major, prepare for shutdown."

"Yes, sir." The Major turned away and spoke into a handheld radio. He was giving instructions to his team to perform an entire shutdown of the electrical services to the building. Some people placed their cell phones, radios, and laptop computers in a large metal box. Several soldiers in the room started to shut down their systems, knowing they would not be needed in the next sixty seconds.

The pilot dove through the clouds and popped out underneath them. He was flying so fast the rain had no impact on his visibility. The screen showing the plane's view instantly went dark, then opened up to show the ground below flying by at supersonic speed. Everything was a blur and barely recognizable.

The seconds ticked away.

The pilot's voice crackled over the speaker. "Red Viper to command."

"Go, Red Viper."

"I am thirty seconds to target. I repeat, thirty seconds to target."

The General looked to a radio operator who shook his head 'no.' "Roger that, Red Viper. Clean Sweep is still a go. You have a green light. I repeat, you have a green light," came the reply.

The pilot responded. "Affirmative."

One of the soldiers at a blackened terminal whispered, "God help us."

The plane roared over the hills and fields, breaking the sound barrier as it tore through the rain just under the clouds, forming a visible cone around the aircraft. The sophistication of the equipment made it almost impossible to miss the target, regardless of where it was, how high or fast the pilot flew, or what the weapon of choice was.

In this case it was a B61 low level thermo nuclear bomb. With 340 kilotons of explosive power, it would utterly destroy any living thing within a two mile radius. Nothing would survive. Nothing.

As the plane approached its target it slowed below mach one. The video transmission of the area became instantly clear. The hills and fields were soaked with rain. A few houses and barns could be seen in the foreground.

A long cylindrical object ejected from the underside of the plane. A parachute immediately deployed, and the object decelerated to less than fifty miles per hour in two seconds. It slowly floated toward the ground.

The people in the room watched the video transmission in horror as an old truck backed out of a barn. The plane kicked in its afterburners and accelerated rapidly breaking the sound barrier as it easily surpassed mach one again and disappeared in the clouds. The video transmission instantly blurred and went dark.

One of the analysts turned to the General. "Sir. Did you see that?" he asked.

"I did. Continue." The General never changed expression.

The cylinder gracefully floated toward the ground, gently swaying side to side in the light breeze. At 200 feet, the bomb detonated.

Year One—2014

Exposure

Saturday, March 16

"ALL RIGHT FOLKS. Listen up." The stocky man spoke quickly but clearly. "Here's the rules of the auction." The crowd of people edged closer to hear the man. "We're going to go inside here and cut the lock off the locker. Then we're gonna open it up. You'll have five minutes to look inside. You cannot go inside. You cannot touch anything inside. You need to stay at the doorway of the locker and look in. As soon as everybody's done lookin', we're gonna sell it to the person with the most cash in their pocket."

A young man working with the auctioneer unlocked the gate as the auctioneer continued. "You'll have twenty-four hours to empty your locker. Alright, are you ready?"

The crowd yelled a collective, "Yeah."

"Then let's go." The auctioneer turned and pointed toward the storage yard as the gate slid open. The crowd of people murmured as they shuffled through the gate and were led by the auctioneer down the rows of storage units. His wife followed close by with a clipboard and a pen. As they approached the first door, the woman flipped to the second sheet on her clipboard. "O.K. We have six units today. This is unit 302b. It's a ten by ten. It was paid for in advance for three years and has not been accessed since 2011 from what the owner of the storage company says." The crowd murmured and several people smiled. One man in the crowd blurted out, "How do they know that?"

"By reviewing the access codes," the lady responded. A few people nudged their partner and smiled as the lady continued. "We'll do this one first to get the blood flowing."

A young man with a handheld grinder started cutting on the lock. Sparks flew as the excitement in the crowd grew. When the lock popped off, the auctioneer slid the door open to the gasp of several people. The unit had several large stacks of white banker's boxes with 'Test Samples— 2002' and other dates and verbiage written in black marker. An oak desk, chair, and floor mat were stacked against the wall. In the back was something covered with a blue mover's blanket. Everything was covered in a layer of dust. A couple of large, black garbage bags lay near the front door.

"That's it?" one buyer said.

Another buyer said, "Looks like someone is going to pay to throw out the garbage."

One buyer laughed as he walked away saying, "You can have their 'samples' if you want."

The crowd was obviously disappointed. People waved their hands in disgust at the unit. Several walked away shaking their heads.

"All right, folks." The auctioneer took control. "I know this looks like whoever is buying this will just be tossing out the garbage, but you never know what treasures may be in there. Who will start the bidding at fifty?" The auctioneer accelerated his speech, blurring his words together as the auction began. "who'lgivemefifty-fifty-fifty-doIhearfifty ...?"

A man about forty years old sheepishly raised his hand. He was wearing a pair of coveralls with grease stains on them.

"IgotfiftydoIhearahundred-onehundred-doIhear ...?" The auctioneer and his staff surveyed the crowd as the auctioneer continued to rattle off a stream of words trying to get the bids. "DoIhearseventyfive-seventyfive ...?" No one raised their hand or bid for the unit except the one man.

"Come on folks. There's gotta be something in there worth seventy-five," the auctioneer's wife pleaded.

"Dollars?" one buyer joked. Everyone laughed.

"SeventyfivedoIhearseventyfivegoingonce ... twice ... fair warning last and final call seventy-five .... sold for fifty dollars!" The auctioneer pointed to the man who placed the one and only bid.

The crowd laughed and clapped as the man smiled and walked over to the locker, closed the door and placed his lock on it.

—*—

An older truck pulling a utility trailer pulled up to the front of locker 302b. Two men stepped out of the truck, unlocked the door of the locker and slid it open.

"That's it? Looks like a bunch of trash here. Bring a burn barrel, Art?"

"You can joke all you want, Elias. We might have something here worthwhile. I only paid fifty dollars for it."

"Yeah. Well, it looks like you could have flushed that fifty dollar bill down the toilet just as easily. At least then we wouldn't have to work on the weekend, too."

"Well, we may as well get this done. Why don't you start with the boxes and I'll get this furniture out?"

"Sure." Elias was apathetic. He would rather be home watching the March Madness basketball tournament than be here digging through someone's trash. But a deal is a deal. They agreed Art would buy the units and Elias would help clean them out for half interest. That way Elias wouldn't have to fork out any money, and Art wouldn't have to do all of the work like he normally did. The problem from Elias's point of view was that Art wouldn't spend much on any unit, so it seemed like they always got crummy ones. "We may as well just throw all of this into the trailer and head for the dump."

"There's some good stuff here," Art said. "I can feel it."

Elias pulled out some boxes and started to rummage through them. "Anything of value in those boxes?" Art asked.

"No. Just a pile of papers about some kind of testing, some notes, some test tubes or something, and lots of these." Elias tossed a plastic bag of wheat kernels at Art. The bag burst open as it hit him, and spilled all over the floor.

"Great. You're just making a bigger mess, Elias."

"Like it matters. Why would anyone store wheat kernels anyway?" he asked.

"Don't know; don't care," Art replied as he dragged the furniture to the trailer making a path to the back of the unit. "Just keep looking."

When he was done with most of the furniture, he went to the back of the unit and pulled a blue moving blanket off an object that it covered, revealing a black fifty-five gallon drum. The drum was sealed and had some labels on the side. "Look at this," he said.

Elias turned from his task of searching boxes and looked at the drum. "All I see is 'Caution.' Great! Now we have some hazardous chemical waste or something to deal with. What is it?"

Art leaned closer to the labels to read them.

CAUTION

Fertilizer 16-MS

Test 12

Expiration date January 31, 2012

Altore Chemicals, Vacaville, California

Ship to Biotrogen

3395 Game Farm Road

Sacramento, CA 95814

"Says it's fertilizer that expired two years ago. It says it's from Altore Chemicals in Vacaville. Being shipped to Biotrogen in Sacramento." He turned to Elias. "Biotrogen? I heard that name before," he said.

"So? It's just a company." Elias dismissed Art's concern with a wave of his hand. "And this unit is just another unit that's going to cost us. Nothing like going in the hole."

Art snapped back. "Would you stop complaining?"

Elias became irritated that he was wasting his day digging through trash and now they might have to pay to dispose of hazardous waste. "Complaining? I haven't even started complaining. You go and buy an obviously trashy unit and now we have to clean it out and maybe pay someone to haul *that* stuff away. What were you thinking?"

Art sat down in a chair sitting against the wall. "I don't know. It was cheap. I thought we might find something."

"We did, partner. We found this trash," Elias shouted as he turned and kicked a banker's box attempting to spew papers all over the storage unit, but the box was solid and didn't move. "Ow!!" Elias fell to the ground and grabbed his foot as he rolled back and forth. "I think I broke my toe!"

Art started to laugh, and then realized Elias was really hurt. "You're serious, aren't you?" he asked.

"Yeah!! I am! I think I broke my toe. What's in that stupid box? Lead?" He slowly raised himself to sit on a stack of boxes as Art walked over to the box, pulled out a pocket knife and cut the tape that was holding the lid on. "What the ...?" The box was full of one-pound coffee cans. He reached in and pulled out one can. He had to use two hands. "This is really heavy." He pulled the plastic lid off of the can, glanced in, and froze. A huge smile appeared on his face.

"What is it? What's so funny?" Elias asked, still holding his foot.

Art leaned toward Elias and tipped the can so he could see in. "Look."

Elias leaned forward and, as he peered into the can, his mouth slowly opened in disbelief. "Gold!"

Art reached in and pulled out a shiny 1996 gold Krugerrand from South Africa. A picture of a gazelle was on the back with the words *"Fyngoud 1 OZ Fine Gold"* inscribed across the bottom. "Looky here," he said as he handed the coin to Elias. He set the can down and pulled out other cans, peeling the lids off each one revealing the gold Krugerrands piled inside. Elias forgot about his sore foot and helped Art empty the box and lay the cans out.

"Look at this!" Elias shouted.

"Quiet!" Art dashed to the front of the locker and glanced out of the unit to make sure no one was around. He went back inside and continued to lay out cans.

"Here's another one," Elias whispered, dragging a banker's box out of the stack. "And here!!!!"

The two men laid the coffee cans and other small containers full of gold on the floor behind a tall stack of boxes. They stood back and stared in awe at the dozens of containers full of gold.

"There's gotta be a million dollars here," Elias whispered.

"It's gotta be stolen or something. Why would anyone keep it here?" Art questioned. "It's gotta be that company: Bio something."

"I don't know. What I do know is we can't tell anyone about this. No one! If this is stolen, I don't want someone to come claim it. It's mine!" Elias smiled as he looked at the fortune.

"You mean, 'Ours.' Right?"

"Yeah. Ours." Elias couldn't take his eyes off of the fortune that lay on the floor in front of him. He just smiled and rubbed his hands together.

Sunday, March 17

THE TWO MEN sat quietly in the truck as it rolled down the country road through the fog toward the storage facility. Elias broke the silence. "You mean to tell me the company, Biotrogen, the company on that barrel, was the same one that had the murders a few years ago?"

"Yeah," Art answered. "I looked it up last night. There were the two owners that got into some kind of fight. One guy killed the other, then committed suicide when the police showed up. He was the guy who killed an employee."

"Yeah, I remember that. It was pretty big news." Elias paused. "Are you sure it was the same company?"

"Positive."

Elias kept driving as he thought about the gold safely hidden away. "Well, I don't think it matters anyway. There is no way we can tell anyone about this, so it doesn't really change anything."

Art disagreed. "We don't know if any of this stuff is related to all of that."

"Still doesn't matter. That was three years ago. It's done, over, kaput." Elias waved his hand in a dismissive fashion. "Old news. We just need to get rid of this stuff and move on." Elias looked at Art and continued. "Besides, you don't plan to return any of that gold, do you?"

Art looked at Elias with disbelief. "Are you kidding? No way!" Both men let out a laugh, and Art's concern quickly vanished.

The early morning fog made the air crisp and cold. The city of Elk Grove was located twenty miles south of Sacramento, nestled in the California San Joaquin Valley; a valley notorious for thick, tule fog. The fog would, at times, be so thick a driver couldn't see the yellow line on a road in front of them. Multi-car pileups were common in the fog-shrouded valley.

"Almost missed it," Elias said as he turned into the driveway of the Storage Place. He pulled up to the gate and punched in the code.

"It's so cold," Art said. "I'll be glad when we get this emptied and done."

"Me, too."

The gate slid open, the pickup and trailer drove through and pulled up to unit 302b. Art unlocked the unit door and slid it open. All that stood in the unit was the large drum full of expired fertilizer.

Art and Elias emptied the unit yesterday, except for the drum. They took all of the papers, test samples and furniture to the dump. They took the gold to Elias's house and locked it in his gun safe where they knew it would be safe for now. Elias was single, so they knew no one would be snooping around his alarmed house. Besides, Art was the one who paid for the unit. Better to keep it away from his place just in case. They counted the Krugerrands last night; 2,122 pieces of South African Krugerrands. They were rich, but still had a very big problem, and it stood right in front of them.

Elias took the dolly out of the trailer and rolled it over to the drum. Art tipped the drum back so Elias could slide the lip of the dolly underneath. "Now." Art let the drum down and rocked it toward Elias until he had the full weight on the dolly.

"This is heavy. Do you think …?"

Elias cut Art off. "No, there's no gold in here. Listen to it slosh." Art put his ear near the drum as Elias shook it. "See. You can hear it's liquid."

"Yeah. But wouldn't it be nice if this had some more?" Art asked.

"Maybe, but I think two thousand ounces is just fine," Elias retorted. "That's almost four million dollars!"

Art laughed. "Can you believe it? Four million!" Both men let out a laugh. "I'm gonna get a ranch in Idaho and hunt every day."

"O.K. O.K. Let's get this done. Then we can go shopping." Elias rolled the drum to the trailer and up the ramp as Art steadied it. They tied it into the back of the trailer, closed the unit door and drove away.

Unit 302b was empty.

Art and Elias were rich.

—*—

The truck and trailer pulled up to Green Creek RV Repair. The gate to the RV yard was closed. It was still early on a Sunday morning. Art stepped out of the truck and unlocked the gate so Elias could drive into the yard. Art closed and locked the gate as Elias pulled the trailer up to the RV dump station.

"This is a great way to get rid of this stuff," Elias said. "It's better than dumping it down the gutter or something. If it got into the storm drains, it could make it to the river and likely kill a bunch of fish, and I don't want to do that."

"Sure. You don't want to do anything like that, but we are still breaking the law, you know." Art was a little irritated that they had to use the RV dump where he works. "If my boss finds out, I could get fired."

"Oh, poor baby. Then you'd have to survive on unemployment benefits and a couple of million dollars." Elias laughed at the absurdity of Art being concerned about losing his job. Art soon joined in.

"I just don't want to get caught, Elias. That's all."

"I agree. So, let's stop talking and start dumping." Elias rolled the barrel over to the RV dump. Art pried the lid off with a crowbar. The black liquid inside reeked of a foul odor.

"Doesn't smell like fertilizer," Art said.

"Like you know what industrial fertilizer smells like?"

"No. But this is really bad." Art looked around to see if anyone was watching them. "I don't like this. What if someone catches us?"

Elias became quickly annoyed. "Listen. It's just going into the sewer system and get all cleaned out. The sewage plant has a bunch of equipment that will sift and filter this stuff until you could drink it, so don't worry. Besides, I don't see anyone here except you and me ..." Elias glanced at the drum and continued, "... and this. Let's get rid of this stuff and go home."

"I guess." Art helped Elias slowly tip the drum into the RV dump pad. The thick, black liquid slowly poured out of the barrel and into the hole in the ground. The odor was horrific. The men made faces and gagged as they emptied the drum down to the last drop.

"That was disgusting!" Elias stepped away from the drum to catch his breath. "Let's get this into the dumpster."

"No. Not here. My boss will see it and ask about it. Let's just take it to the scrap yard. They'll take it if it's empty. We just have to rinse it out."

Elias laid the drum onto its side as Art grabbed the hose. The men washed out the inside of the drum and returned it to the trailer and pre-pared it for the trip to the scrap yard.

Elias and Art were unaware that they had just released a toxic chemical that, three years earlier, was responsible for the deaths of nearly 60,000 people and the ultimate bombing of Quincy, Washington.

The thick sludge mixed with the wastewater in the sewer system and started its long journey through the underground pipes and tunnels to its new home: the Elk Grove Wastewater Treatment Plant where billions of bacteria and viruses were awaiting it's arrival.

Monday, March 18

THE ELK GROVE Wastewater Treatment Plant was known for its integration of a sophisticated wastewater treatment facility into a suburban community environment. Its cutting-edge process of wastewater treatment and reclamation was a model for future plant designs. The plant was located south of the city between the Consumnes River and the Applegate Wetlands. It covered seventy-five acres of prime farmland with pipes, tanks, brick buildings and cement ponds.

The facility was enclosed with a seven foot chain-link fence topped with six strands of doubled barbed wires strung between "V" shaped caps. It looked more like the fence of a prison yard than a wastewater treatment plant. Large Oleander bushes attempted to cover the ominous fence. The entrance gate had several cameras focused on the driveway and nearby street. The Operations Building sat just inside the gate. The brick building with large windows was trimmed with colorful rose bushes, juniper and camellia bushes. A row of yellow marigolds bordered the slate walkway to the front entrance. The scent of fresh, redwood bark filled the air.

Quinton Jacob Lemolo sipped his coffee as he walked into the lobby of the Operations Building. "Mornin'," he whispered in a raspy Texan voice as he passed by the reception desk and headed for the restroom.

"Hi, Quint," came the reply from Grace, a lady in her late fifties with dark hair and black rimmed glasses connected to a chain around her neck. "You look and sound terrible! Are you O.K.?" she asked.

"Yeah. I reckon," he said with a slight drawl as he continued down the hall. "Be right back," he yelled as he stepped into the restroom.

After a few minutes, he emerged, took a few steps down the hall toward Grace, stopped, quickly turned around and went back into the restroom.

"Poor boy," Grace whispered as she went back to work at the counter. Soon, Quinton came out again and slowly walked up to the counter blowing his nose with his handkerchief. "Ya know, just when you think you have the horse in the barn, it takes off running for the pasture again." Grace laughed as Quinton wiped his nose. "I got this bug Saturday. It's murder," he said as he blew his nose again. "Nothin' seems to work, Grace. I took some anti-histamine, but I still feel lousy. My throat is killing me. And you saw what else is ailing me a minute ago."

"Why don't you just turn right around and head back home, young man?" Grace said in a motherly tone. "We don't need you here spreading your germs all over the place."

"Grace, I would, but James is not coming in this week. He's on vacation so we're down two guys." Quinton wiped his nose again. "I just need to check the systems and then I can take a nap or something."

"Maybe you should go see a doctor."

"I don't like doctors. I'll get over it."

"O.K. But I'm warning you, young man." Grace pointed her finger at him and shook it. "You better not get me sick." She tried to contain her smile and appear stern, but couldn't. She liked Quinton. Everyone liked Quinton.

Quinton joined in the banter by raising his hand as if he was asking her to stop. "O.K. I promise I'll lay low. Use some wipes on the phones and stuff." He turned and walked into the main operations office next to the reception area.

Command central for the facility looked like a typical office. Calendars, pictures of cars and photographs were either pinned or taped onto the walls and partitions throughout the area. Piles of papers littered desks and tables. Manuals were crammed into shelves and stacked in the corners. A long workbench had seven monitors lined side-by-side displaying a variety of colored shapes, graphs, charts, tables and diagrams. Each monitor reported on a specific function or area of operation in the entire facility. With five thousand sensors and monitoring points, the slightest variation of any activity would result in an alarm.

"Hmmm." Quinton scanned the panel of lights and monitors across the station. He punched a few keys as he finished his coffee and sat it on the table next to the station. He punched a few more keys and took his jacket off. "Great. We've got us an event." He sat in the chair and began scanning through the various screens to identify what happened when and how to resolve it.

Grace peered into the room. "Would you like some hot tea?"

"Huh?" Quinton turned to face her. His eyes were swollen and red. He was a young, twenty-seven year-old man in great physical shape, except for the cold that was beating him up. "Tea? Well, if you don't have black coffee boiled over a campfire then, yeah, that sounds pretty good right now." He flashed a big smile thankful for the concern that Grace showed.

"Picky. Does Abbey have this cold, too?" Grace asked.

"No. She's fine," he said.

"Good. You don't want to get your soon-to-be mama sick if you can help it." Quinton was a newlywed, married only six months. They were expecting a baby in just five months.

"I know. I'm being as careful as a long-tailed cat in a room of rockin' chairs." Quinton blew his nose and coughed.

Grace smiled as she exited the room. Quinton turned back to the monitors and the problem at hand. He had worked at the facility for nearly six years and knew a lot about the operations. Many people thought he was uneducated because of his drawl. He would often play on the stereotype just to confuse them, when in fact, he was quite brilliant. His degree in environmental studies coupled with his interest in biology enabled him to quickly grasp the operations of the plant.

The wastewater from the metropolitan area was pumped by several substations to the facility. It started its cleansing journey at the top of a five story brick building at the back of the property. Five large screw pumps would pull the wastewater from the six foot wide main incoming line to the top of the building where it would continue its journey through the entire facility with the help of a gravity-fed system. First, heavy, large objects were screened from the wastewater and compacted into removable debris. Rags, sticks, paper, rats, dismembered bodies; anything that ended up in the sewer system would be screened, ground, compacted, and disposed of in large, two story dump trucks. One truck a week would make the trip to the dump.

From there, the liquid flowed to four large aeration clarifiers that looked remarkably like large oil tanks. Elements heavier than water would settle to the bottom of the tank and were pumped to anaerobic digesters to be processed into "biosolids." The biosolids were then pumped offsite to the Elk Grove Biosolid Management Farm located five miles away near a hop ranch. There, they were turned into beneficial soil amenities by filtering underground through a large grove of poplar trees.

The anaerobic digesters created a huge amount of methane gas as the bacteria would attack the debris in the water causing the gas to release. The methane was sucked off the clarifiers into a generator building and used to power the majority of the operation. Excess methane was burned through an eternal flame at the rear of the powerhouse. There was a rumor that Quinton had, at one time, used the lift truck to elevate himself to an area near the flame and roast some hot dogs for dinner. Of course, it was just a rumor.

The wastewater continued its path through the facility with its next stop at the aeration basins. These were large, rectangular concrete ponds that looked very similar to those found at a fish hatchery. A two-story building next to the aeration basins housed six monstrous pumps that sucked air into the pumps and carried it through the pipes to the bottom of each pond. Bubbles rose from the pipes and agitated the wastewater causing the chemicals in the wastewater to separate, creating a type of scum called "floc" that floated on the surface in cloudy clumps. "Good bacteria" was brought into the wastewater to mix with oxygen in a biological process to dissolve and absorb the remaining organic matter in the water.

The floc was removed as the water flowed into secondary clarifiers, which were eight large open tanks with water sprayers on top. The good bacteria and matter was settled out and recycled into the plant to do its work all over again. The water tanks were an inviting pond for the local ducks, sea gulls and blackbirds. The birds swim in the effluent and eat the floating pieces of leftover floc and scum loaded with complex carbons and nutrients. Quinton swore he would never eat waterfowl for dinner again after seeing them eat the scum from the ponds.

The cleaned water rose to the top and overflowed into the final concrete channel taking it through the disinfection process. Chlorine, hydrochloride acid and other chemicals were introduced to the water at the entrance of the channel to disinfect the remaining bacteria. The chlorine and harmful chemicals were then removed near the end of the channel through a complex series of filters. The water continued its journey and was eventually discharged into the constructed wetlands adjoining the natural wetlands. Samples were automatically drawn out of the flowing water at four points along the way by automatic suction stations and pumped into an "automated lab" where it was analyzed for bacterial and chemical content.

This was where the "event" was occurring.

Quinton could easily tell by the monitors that the automatic sampler was not grabbing the right number and volume of samples. "Looks like a plugged valve," he mumbled. He grabbed a radio, his keys and a hat and started to the door when Grace stopped him. "Here. This will help you feel better." Grace handed him a large cup of steaming tea. Quinton took the cup and sipped some. "Oh. That's mighty fine tea, Grace. What is it?"

"Lemon honey 'Throat Coat'," Grace said. "It will help clear up your nose and sooth your sore throat."

Quinton took a deep breath through his nose. "It is already. Thanks, Grace ... for watching out for me." He took another sip and tilted his head back, closed his eyes and smiled. "Ahhh."

"Glad you like it," Grace said.

"I do." Quinton nodded toward the monitors. "I'm heading out to the lab. I think I have a plugged line or something," he said.

"Okay. I'll hold the fort down ... like I always do," Grace smiled.

The young man grabbed his hat and pulled it over his bushy, blond hair. He held the hot tea in both hands as he headed out the door to his golf cart for a ride to the automated lab.

The lab was at the far end of the plant at the end of two long, concrete channels that ran from the secondary clarifiers to the discharge vent. It was a two-story brick building with no windows. Quinton unlocked the door and entered a large room. A young woman's voice was speaking over the noise of the pumps.

"This is where …" She stopped. "Oh. Quinton." She was surprised by his entrance. "Children, this is Mr. Quinton Lemolo, Assistant Operations Supervisor of the facility." She gestured toward Quinton and smiled. She was attracted to the young man who had a beautiful smile, blond hair, southern manners and a great build. She wondered if he ever felt the same toward her. She didn't care if he was married.

Quinton tipped his cap and waved. "Howdy, there." He sat his cup of tea on a small table next to a device that looked oddly similar to a plastic Christmas tree. It was a two foot tall glass cylinder that had alternating tubes entering the sides at forty-five degree angles every inch or two, stacked the entire twenty-four inches. The device had a digital display on the front that revealed a series of numbers alternating every two seconds as small amounts of water would flush through the alternating tubes to be analyzed. Clinton visually traced the tubes up through the top of the chamber to twenty-four small pipes, each winding their way along the ceiling to incoming points around the room. This was one of twelve chemical analysis monitors in the room. "Looks like a tour," he said as he glanced over to the children.

A young man about Quinton's age spoke up. "Yes, sir. We are half of a third grade class from Elk Grove Elementary," he said as he gestured at the sixteen students surrounding him. "I'm Daniel Wu." The man extended his hand to Quinton.

"Uh. I better not," Quinton said as he looked at the man's hand. "I have a pretty nasty cold or something, so not to be rude or anything …"

Daniel withdrew his hand. "Oh. No problem. Thanks for being careful." He looked around the room. "I can see 'careful' is what this whole process is about," he said as he eyed the surroundings.

"Well, Mark Twain said 'Better to be careful a hundred times than dead once'."

The young woman interrupted. "We _are_ … very careful," she said. "Everything we do here is to make sure the water is cleaned, disinfected and safe before we discharge it into the adjoining wetlands."

One of the girls in the class shouted, "Stop it!" as she swung at a young boy standing next to her. The boy jumped back and raised his arm away from the girl. His arm caught one of the tubes going into an analyzer, and pulled it loose. A small spray of water shot from the tube in spurts. It reacted like the boy had just severed an artery in a person, flopping around uncontrolled and squirting water in every direction.

"Lonnie!" the teacher yelled.

The water kept spurting and sprayed a small amount of water a few feet away onto a little girl standing at the back of the group, watching the drama. It hit her in the face. "Ewww," she yelled as she ducked and tried to avoid another spurt. She frantically started wiping off her face.

The boy ran across the room as Quinton grabbed the spurting tube and pinched the area just above the leak. He reached over with the other hand and shut the analyzer off. "It's O.K. It's just water we're testing before it's discharged." He looked at the young girl who was almost in tears and knelt down to her eye level. "It's O.K., honey," he said. His voice was calm and soothing. He wiped the water away from her face with his hand and shirtsleeve. "It's water that has been all cleaned and is ready to go into the pond with the ducks, so don't you worry about it."

"O.K." she said sheepishly, still upset about being sprayed.

Quinton flashed his big smile. "What's your name, sweetie?"

The girl cautiously glanced up. "Bonnie."

"That's such a pretty name. You don't need to worry about this water, Miss Bonnie," Quinton said as he finished wiping her cheek.

"O.K."

"I'm really sorry about this," the teacher said. He turned to the young boy who instigated the event. "Lonnie, you need to be more careful. There are places to play like a child, and places to behave like a man. This is a place to behave."

"Yes, sir." The boy looked down as he moved closer to the female guide seeking protection from her.

"I'm sorry about this," Daniel said to Quinton and the guide.

Quinton immediately responded. "Don't you worry about it. No use crying over spilt ... water. Won't take me but a few minutes to clean this up. You enjoy the rest of your tour," he said as he nodded at the young woman.

"O.K. Next, we will see where the water is being discharged and then take a walk through the wetlands." The lady clapped her hands and opened the door to the hallway. She looked back at Quinton, smiled and mouthed, "Thank you." Quinton smiled back, slightly tipped his hat and turned to the analyzer as the group left the room.

He coughed a couple of times. Then, he reached over and took a big gulp of his warm tea, which recently received a few drops of the water from the analyzer leak. "Ick. It's getting cold." He took another gulp and tossed the rest onto the floor near the drain and started working on the analyzer tubes.

Tuesday, March 19

GOLDEN POND MEMORY Care Facility was having its open house this week. Several times a year the facility would run full page newspaper ads touting the benefits of personalized care in a relaxed, scenic community. The facility was south of Elk Grove and offered fantastic views of the Consumnes River and adjacent wetlands. Large flocks of Canadian geese could be seen cupping their wings and floating into the marshes for a landing. Ducks swam in the numerous ponds in the wetlands while blackbirds with bright red underwings shrieked from the cattails. The wooden walkway of the wetlands was connected to a high gate that protected a sidewalk from the facility. The residents and staff had multiple opportunities to go on pleasure walks through the marsh just by entering the code into the electronic lock on the gate and exiting the facility grounds into the marsh. Benches, information kiosks, water fountains and telescopes offered vast encounters with wildlife in a serene and controlled environment.

The front of the facility was inviting, with a white exterior and forest green trim. High white fences encircled the side and back of the property making it look like a private mansion. The entryway was large and well decorated with Early American furniture. The foyer and hallways had sofas, chairs, loveseats and end tables scattered about. Various paintings and antiques were hung on the walls. The first name and a picture of the resident were posted at the front of each door down the hallways.

The smell of urine and feces was well controlled today. Some days the odor was overwhelming. The incontinence of most of the residents was an ongoing task for all of staff to manage. But today, staff worked hard to get the odor under control and well masked for the open house.

Melanie Grimes was the administrator and proud of her facility. She was a natural. She cared very much for each resident and staff member. She knew the importance of helping residents and families through the ravages of Alzheimer's Disease and other dementia related illnesses. She knew, because she experienced it first hand with her parents.

"Thank you, Donna Mae." Melanie smiled as she reached out and patted the hand of an elderly lady in a wheelchair. "I will forever treasure these," she said as she took the silk flowers from the woman. The elderly woman watched intently as Melanie raised the flowers to her nose. Melanie smiled as she pretended to take in their fragrance. The elderly woman smiled too, then reached out to take the flowers back. Melanie handed them back to her. The elderly lady took the flowers, smelled them and handed them back to Melanie. Melanie repeated her motions, comments, and gestures as she had done many times before with Donna. The disease in Donna's brain would cause her to repeat the same, simple motion of giving and receiving the flowers a thousand times in a day. That was her life. Nothing more; nothing less. Just the flowers. It was the only way she could connect with someone and Melanie knew it.

The crowd around the two women chatted, sipped on sparkling cider and ate hors d'oeuvres and chips. The room was large, much like a grand ballroom on a cruise ship with a hanging chandelier, paintings on the walls, white lined tablecloths and Early American chairs and décor. The room was abuzz with conversation and activity when a man quickly rushed into the room. "Ms. Grimes?" The middle-aged man in a green uniform worked his way through the crowd and approached the woman. "May I speak with you for a moment?"

"Sure, Adam." Melanie turned away from the elderly woman, cutting off her interaction with her to give Adam her full attention. She could tell something was bothering him. "What is it?"

Adam cleared his throat. "Well, we're almost out of sparkling cider," he whispered.

"Well, the humanity of it all! Break out the champagne, then." Melanie reached over and took a plastic glass off a table nearby and raised it. "Here's to ya," she said as she downed the glass of cider, wiped her mouth with her shirtsleeve and smacked her lips. "Ahhh."

Adam's face turned downcast at the mockery. "Not funny."

"Adam." Melanie leaned closer to her nurse and whispered. "Lighten up. It's O.K. Our guests need to see us handle all situations with a smile and professionalism, even a shortage of cider. So relax, enjoy, and work through it."

"What should I do?" he asked.

Melanie chuckled. "Order more. Send Beverly to get it. Use the credit card. You can figure it out." Melanie turned to greet several guests who had just walked into the cafeteria. Adam turned and walked out, passing an elderly man shuffling into the room with a walker. The man had his head down, eyes fixed on the floor, oblivious to the surroundings. He was wearing blue sweatpants and a red flannel shirt. His stocky frame was hunched over, leaning on his walker. His gray hair was in stark contrast to his black skin. A young aide walked over to greet him as he entered the doorway.

"Hi, Mac." The girl smiled widely and placed her hand on his shoulder, but the man never looked up. He stopped as she touched him. "How are you today?" she asked.

The man stared at the floor. The girl lifted her hand off his shoulder and he started to shuffle away. She touched his shoulder again. The man stopped, still looking down.

"Mac, are you going to sing for me today?" the aide asked.

"Sing?" he asked without raising his head.

"Yes. Please sing your father's song for me. It's so beautiful," she said.

The old man raised his head to look at her. His eyes lit up. He got a smile across his wrinkled, unshaven face as he said, "Oh. My *Father's* song." His posture changed as he straightened and stood tall. His well-built frame looked like a young, confident man instead of the old, timid patient who walked into the room. His voice became resonant as he loudly said, "Madam, I would consider it an honor to sing for you and your guests." He made a sweeping motion of his arm around the room as though he was throwing back his robe, turned and faced the crowd. He cleared his throat.

"Our Father, Who art in heaven." His magnificent voice echoed through the room and down the halls. Everyone in the room became instantly quiet. *"Hallowed be thy name. Thy kingdom come. Thy will be done on earth as it is in heaven."*

The guests in the room were stunned. What a beautiful voice. Had this man sung opera in the past? All eyes were fixed on the old man who was standing and singing like a young man, full of energy and life. He was a new person.

"Give us this day our daily bread, and forgive us our debts, as we forgive our debtors."

The patients in the hall outside the room began gathering in front of the door and peering in to see where the captivating voice was coming from. They were shuffling into a small group and gently pushing the people in the front of the crowd into the room. The group slowly spilled into the room.

"And lead us not into temptation, but deliver us from evil."

He took a deep breath to fill his lungs for the grand finale. He proceeded with clarity, confidence and power.

"... For Thine is the kingdom ...." He stretched out his arms and arched his back for the full delivery of the song. His eyes were looking to the ceiling as though he was singing directly to heaven itself. Everyone in the room was in awe of the beautiful, powerful voice that was singing to the heavens.

"......... and the power, and the glory forever. Amen."

The room exploded with applause. Several people shouted "ENCORE, ENCORE." The old gentleman stood with his arms fully outstretched, then bowed low with a sweeping gesture of his hand. As he slowly rose and scanned the pleased audience, he had a tremendous smile across his face. Then, the smile quickly faded and he resumed his old, bent posture leaning on the walker. He shuffled through the applauding crowd, out the door and down the hall, disappearing into the mix of other lost patients.

A couple standing next to Melanie clapped heartily. "That was quite a performance," the woman commented.

"Yes, it was," Melanie replied, clapping with the audience. "That was Mr. Stanley MacKenna Thorne. We call him 'Mac'."

The applause settled down and the crowd returned to their mingling, most commenting on the outstanding performance they had just observed.

"Was he an opera singer or something?" the man asked Melanie.

Melanie smiled. "Yes. Or something would be appropriate. Mac was a young man at Pearl Harbor when it was attacked. He served in World War Two, received the Congressional Medal of Honor, married, had a bushel full of children, ran a successful business, started the local mission and eventually became a world class tenor."

The woman was shocked. "Oh, my! That is really something. What a story! He sings so beautifully."

"Yes, he does." Melanie smiled and extended her hand. "I'm Melanie Grimes, the administrator here."

The lady took her hand. "Janice Powell and my husband Greg."

"Welcome to Golden Pond. I trust you are enjoying the open house?" Melanie asked.

"We are, thank you," Greg replied. "I like the name Golden Pond. Is it named after the movie?"

Melanie smiled. "No. It's named after the 'pond' that we are next to. The Applegate Wetlands adjoins this property, and we have a very large pond right out back with a wooden walkway that connects to our facility."

Janice was concerned. "Seems like the residents could end up drowning if they went there."

"Oh, no. We have a very secure campus. No one goes to the pond unless they are escorted by someone with a pass card and the code." Janice smiled. "Our aides take residents there regularly, one on one, to have a break from the surroundings here and a little external stimulus. It is a part of our care and therapy here."

"That is very interesting. We are looking for a nice facility for my mom that really tries to do activities with residents and cares for them. She has Alzheimer's that is progressing rapidly, so we need to move her to a secure memory care facility. We can't take care of her any longer."

"I'm sorry to hear that," Melanie replied. "My father had Alzheimer's, so I fully understand. It is a difficult disease to manage."

The woman sighed. "I never imagined a person could be so damaged by it. Mom is deteriorating so quickly." The woman got tears in her eyes as she spoke.

Melanie placed her hand on her shoulder. "It is a terrible disease. But there can be moments of joy, like we just saw with Mac. He has Alzheimer's and doesn't remember anything about his past, his heroics, family, or abilities. But he can connect with one thing: that song. He can't sing anything else because he doesn't remember it. Only that song. So, we let him sing."

"It was wonderful," Greg said.

"Yes, it was. Let me show you around the place, if you have a moment," Melanie offered.

"Sure," Janice replied.

The three of them walked out of the room and started down a hallway. "This is our entry level wing. We just left the main dining room. This wing is where the lower level patients live."

"Very nice," Janice said as they passed several rooms with pictures of residents at each door. "Oh, here's Mac." Janice pointed to the picture.

"Yes," Melanie replied. "Mac lives here because he is very functional with his daily living tasks; bathing, dressing, toileting. He can take care of himself with minimal assistance and prompts, and he's mobile. That makes it a lot easier for us to manage his care."

"What happens as the disease progresses?" Greg asked.

Melanie paused. "Are you aware of how the disease progresses?"

Janice looked down to the floor and replied. "We are, somewhat. Mom's doctor explained quite a bit and gave us some brochures."

Melanie spoke slowly and intentionally. "Most people I have met have an awareness of what the disease does, but really do not know. Most doctors sugar-coat it to some degree, explaining some of the behaviors that are demonstrated, but they don't tell the real story."

"What is the real story, if I may ask?" Greg asked.

"I don't want to shock you. It is a terrible disease. If you have a few minutes, I can show you some of the advanced patients in the next wing. Reading about it and seeing it is totally different," Melanie said.

"Sure. We know it's not good. Anything that will help us understand what lies ahead has to be helpful," Greg replied.

"I don't think I want to see," Janice said. She looked at Greg. "Can I just stay here?" she pleaded.

"Sure, hon." Greg gently touched her arm. "I'll be right back." Greg and Melanie walked over to two large double doors. Melanie entered a code, the door buzzed, they opened it and walked through. The door quickly latched behind them.

Janice sat on a loveseat facing the window. She could see the wetlands outside and a wooden walkway that meandered through the cattails and reeds. A pair of Mallard ducks was coming in for a landing right over the top of an aide pushing a man in a wheelchair toward the gate. She recognized the man. It was Mac; the man who just sang for the group. *What a beautiful place*, she thought.

—*—

The smell of urine and ammonia was powerful, much stronger than in the other wing of the facility. Greg looked at the residents scattered along the hallway walls in their wheelchairs and walkers. One elderly woman near them was sitting in her wheelchair methodically flipping the pages of a magazine. The magazine was upside down. "Does she realize that the magazine ...?"

Melanie interrupted him. "No. She doesn't know what it is. She is attracted to the colors, shapes, sound of the paper turning and the rhythm."

"That's sad," Greg said.

Melanie agreed. "It is. Alzheimer's is one of the worst diseases I can think of." Melanie leaned down to the woman and slowly lifted her chin to eye level. She kept her fingers under her chin, smiled and softly said, "Hello, Miriam." The lady smiled, but never said a word. As soon as Melanie removed her fingers, the lady stopped smiling, lowered her head and began methodically flipping the pages of the magazine again.

Greg was stunned. He observed the odd behaviors of the other residents. A man was pushing a woman in a wheelchair toward the hallway wall. A few seconds later, they turned around to go back from where they came. Then, they turned again; over and over, back and forth. A man standing at a door, opened it; then closed it. Again and again; over and over.

Melanie let Greg take the scene in for a few seconds. "Repetition is soothing for them, Greg."

"Huh?"

"Repetition. The disease robs people of memory and the cognitive ability to process information. As their condition worsens, their world shrinks considerably; from a town to a house to a room to a chair to ..." Melanie glanced to the woman with the magazine and continued "... to a magazine."

Greg was almost sick. He could feel the nausea beginning to overwhelm him. "It's horrible." The smell of urine, the blank faces, the lights were all becoming too much. "I think I need a drink of water or something."

"Here. Let's go in here for a second and catch our breath." Melanie opened the door to a small room with a vending machine, table, some magazines and a television. "This is the employee lounge for this wing."

Greg quickly sat at a table and put his head in his hands.

"Are you O.K.?" Melanie asked as she handed him a glass of water.

Greg looked up to Melanie. "Yeah. I'm feeling better already. Thanks." Greg took a sip of water. "I had no idea how bad this disease is."

"It is truly terrible, but fascinating at the same time."

"How do you mean?"

"A patient will have no recollection of who a son or daughter is standing next to them and then, all of a sudden, remember everything for a few seconds or minutes. They will know who they are, their name, grandkids, everything. They temporarily come out of a fog only to return."

"That's horrible!"

"It is, yet if a family member can experience that, it is a joy to reconnect for even a few seconds. The mind is an amazing part of our body that we really know very little about."

"I had no idea. Is there another level beyond this one?"

"This is the highest level of care next to hospice."

"Hospice?" Greg asked.

"Yes. Hospice is the dying stage. They are a team who help the resident and family members through the dying process. Most people diagnosed with Alzheimer's live about seven years before they die from the disease, depending on the age they are diagnosed. Some live much longer."

"Die from it?"

"Yes. Severe dementia, which includes Alzheimer's, frequently causes such complications as immobility, swallowing disorders and malnutrition. These complications can significantly increase the risk of developing pneumonia or dehydration, which has been found in several studies to be the most commonly identified cause of death among elderly people with Alzheimer's disease and other dementias."

"How does Alzheimer's cause pneumonia or dehydration?" he asked.

"Alzheimer's attacks the hippocampus of the brain; the area that manages memory. The dead and damaged cells are unable to connect properly, causing shorts in the brain's electrical impulses. The person loses more and more memory function until they basically forget everything including body functions. They will forget to eat or drink or to swallow. They forget the basic elements of how to live." Greg was impressed with her knowledge and relieved he was out of the 'hallway of despair.' He was enjoying the educational lesson from the knowledgeable woman. "The situation has been described as a 'blurred distinction' between death with dementia and death *from* dementia."

"I can see it would be difficult to sort out." Greg stood to his feet. "What a terrible disease."

"It is. You look like you are feeling better."

"I'm feeling much better, thank you." Greg glanced toward the door. "After seeing the condition of some of these people, I have absolutely nothing to ever complain about. Nothing."

Melanie smiled. "We can certainly gain a new perspective when we encounter something as traumatic as this."

"Yes, we can," he said. "Shall we continue the tour?"

"Are you sure?"

Greg didn't hesitate. "Absolutely."

—*—

Janice was sitting on the loveseat daydreaming, staring out the window, when the doors leading outside to the walkway burst open and the aide rushed Mac into the lobby in his wheelchair. Mac was shivering and coughing terribly. "Oh … what happened?" Janice asked as she jumped to her feet.

"He fell into the pond," the aide said as she wheeled Mac past Janice and down the hall. Water dripped from Mac's soaked pajamas and robe.

"Fell into the pond? How did that happen?" Janice asked.

The aide never looked back; never responded. Mac was coughing as the aide quickly pushed him down the hallway into his room. Janice watched in disbelief as the two disappeared into the room with Mac's picture next to the door.

Changes

May
Two months after exposure

QUINTON LEMOLO WAS sitting at the monitoring station of the Elk Grove Wastewater Treatment facility reading a book when Grace walked in.

"What are you reading?" she asked.

"Oh. It's a compilation of various literary works called Norton's Anthology of American Literature." Quinton's drawl barely came through as he spoke. He held the book up showing its voluminous size. "It's seventeen hundred and sixty pages."

"Are you serious?" Grace was appalled at the sheer size of the book. "What on earth is in that thing? A full set of encyclopedias?"

Quinton chuckled. "No. Not hardly. Though it looks like it could be ..." he said as he turned the book over in his hands. "It's a collection of writings by American authors from the inception of our country through modern history. This is volume two."

"Volume two?" Grace was stunned. "You mean you've already finished the first volume?"

"Yes, ma'am. Last week," Quinton replied matter-of-factly. "I've got this insatiable desire to read. The words just jump off the page at me. It's interesting, informative and educational. I've read the works of Shakespeare, Dickens, some history ... did you know Lewis and Clark only knew each other for six months before they joined together in their adventure across the continent?"

"No, I didn't," Grace replied.

Quinton continued. "And Sacagawea was only sixteen years old and gave birth on the journey."

"I heard something about that." Grace paused. "Quint, are you O.K.?" Grace's concern showed.

"Okay? I'm grrreeeaaattt!" he said, mimicking the Tony the Tiger cereal mascot.

Grace did not see the humor in his answer. She believed something was wrong. "Seems like something is ... is going on with you."

"What do you mean?" he asked.

Grace was wringing her hands because of the uncomfortable conversation, but she was concerned about him. "Well, you're reading all the time, talking really fast, you sound different, and you seem almost bored at your job."

"No. Reading is good, Grace. This is fascinating stuff. You can see the development of American literature over time if you read through the volumes without interruption. You get a good picture of the development and metamorphosis of American literature and culture represented ..."

"Quint!" Grace yelled. "That's what I mean!"

"What?" Quinton looked up from his book with puzzlement. "What are you talking about?"

"Stop talking like a professor of literature or something." Grace stepped closer to her friend. "I'm concerned something is wrong with you. You never read anything here before. Not even a magazine. Now, well, look at you. And you're using words like I've never heard you use before. Metamorphosis?"

"It means change."

"I know what it means!" Grace replied loudly.

Quinton never changed expression. "You don't have to shout, Grace. There's nothing wrong with me." Quinton closed the book. "I've just developed a fascination for reading. These books are great."

"What about your job? Do you even know what's going on in this facility?"

"Of course I know, Grace. I wouldn't abdicate my responsibilities."

"Are you even watching the alarms or monitors to see …?"

Quinton cut her off. He never changed expression as he stared at her and without looking at the monitors, said, "There's a level four alarm at station seventeen indicating increased pressure in line two, a 'B' alarm at valve twenty-four in Primary Clarifier two, surface aerator six is at 1.935, aeration basin four is at 12.4 degrees Celsius, PH in discharge tube three is down to …"

Grace glanced at the monitors and saw that Quinton was listing every alarm and indicator on screen five, the alarm screen, in sequence. "Stop!" Grace shouted. "Stop it!"

Quinton blinked his eyes. "Stop what?"

Grace turned and quickly walked out of the room.

Quinton watched her exit and shrugged his shoulders. "Strange," he mumbled and returned to his book.

—*—

Daniel Wu was sitting at his desk in his third grade classroom at Elk Grove Elementary School. It was late Friday afternoon. He was a young, ambitious teacher ready to educate the world. He loved creating visual representations of topics that the class was studying to use as a constant reminder and reinforcement of the topic materials. He just finished removing some poster board materials and displays depicting the process of photosynthesis. He planned to replace the display with the materials representing the levels of atmosphere surrounding the earth while his aide finished grading the papers from the last test.

Pamela Crawford was sitting at the table with two stacks of papers and a laptop computer in front of her. She was like the grandmother of the class. Pamela was in her mid-fifties with reddish blond hair. She was a joy to have and a God-send for Daniel. "Mr. Wu?"

"Yes, Pam."

Pam was looking at a paper in her hand as she spoke. "Have you noticed the grades Bonnie Howe has been getting on her tests?"

"Sure. She's improved quite a bit," he said as he pulled some materials out of a large bag. "I think she even got an A on her last test, didn't she?" He held up a huge picture of the earth and positioned it on the wall with some tape.

"She aced it," Pam replied. "In fact, she has aced her last two tests."

Daniel stopped and glanced over his shoulder. "Both of them?"

"Yes. Both of them."

"Even the last one on math?"

"Yes. Math and science ... both of them." Pam slowly placed the paper on the stack. "Do you think she is cheating?"

Daniel turned around and the large earth fell to the ground. "Cheating. No! Not Bonnie. She's having enough difficulties with her medical condition. She sits in the front row. No. No cheating there." Daniel walked over to the aide to look at the latest test.

Pam handed him the paper. "See? And she was the first person to finish the test."

"So, you are watching her and still suspect she is cheating?" he asked. He looked over the test.

"How can she be doing so good when she was struggling so much just a couple of months ago?"

"I don't know, but she is doing *well*," he said as he smiled at the aide. "This is a perfect test. I'm not sure I've ever had anyone get a perfect score on this test. And I've been teaching this class for four years."

"See? Doesn't that say something?" Pam asked.

Daniel thought for a few moments, then replied. "I think what it says is that she may be a lot smarter than we gave her credit for, or she has found a way to understand the material, or something."

"Yeah, or something," Pam said sarcastically.

"Let's try something. Let's administer an IQ test to the kids next week. We can do it in three parts for about fifteen minutes each."

"What test?" Pam asked.

"It's a test I found online. I've used it before. It's free and simple. It was developed by Doctor James."

"You mean Doctor James from TV?" Pam asked. "That Doctor James?"

"Yes. It's a reputable test and fun. It can give us some idea of where the kids stand, particularly Bonnie. If we see something abnormal, we can decide how to approach it. It's worth a try."

"OK. I like it," Pam agreed. "Will you get it together or do I need to do something?"

"The only thing I need to have you do is finish the grading ... and help me with this layout." Daniel pointed to the large earth still lying on the floor.

Pam snickered. "At least the sky isn't falling."

—*—

Stanley Thorne was staring at the image in the mirror when an aide walked up behind him. He didn't look at her. He was captivated by the man staring back at him. He looked disheveled. His gray hair was uncombed and there was beard stubble all over his wrinkled face. The man in the mirror was wearing pajama bottoms and a t-shirt.

An aide walked into the room carrying some toilet paper. "How are you today, Mac?" she asked, never expecting an answer, because Mac never spoke, except when he sang.

"O.K., I guess," he replied.

The aide dropped the toilet paper and froze. She turned and stared at the man standing in front of the mirror. "Mac? Did you just say something?"

Mac turned toward the young woman. "Do you need hearing aids or something? I said I was O.K., I guess." He turned back to the mirror and continued to stare at the man. "Is that me?" he asked.

The girl quickly moved to his side. "Yes, Mac," she said as she placed her hand on his shoulder and stood behind him looking into the mirror. "That's you."

"Damn. I got old."

"We all get old, Mac, if we live long enough."

Mac turned to the girl and continued. "I don't remember me looking like this." He turned back to the mirror. "How old am I?"

"How old do you think you are, Mac?" the girl asked.

Mac got irritated. "Can't you just answer the stupid question?" He turned to face the girl without the use of his walker, which was sitting across the room near the bed. "I'm not playing a game here. I don't remember me looking this old and today, I get up, go to the bathroom, look in the mirror and see an old man. I just want to know how old I am."

"O.K., Mac. It's O.K. You're ninety-one."

Mac turned back to the mirror and stared. "Ninety-one. Ninety-one." He lowered his head.

The girl was shocked. "Mac. I think you should sit down for a minute. I need to get the nurse and have him look at you." She gently placed her hand on Mac's arm. "Right over here, Mac. Let's see if we can figure out what's going on, O.K.?"

Mac let the girl escort him to the side of the bed where he sat down. He was looking at the back of his hands. "What happened to all those years?" he asked.

"I don't know, but let's get someone to help us sort this out. Can you sit right here for a few minutes while I get the nurse?"

"Sure. I'm not going anywhere," he said as he continued to inspect his hands and arms.

The aide hurriedly left the room.

Mac continued to examine his arms and suddenly became downcast. "Mabel," he whispered. His eyes quickly filled with tears. "Mabel."

Mac sat on the edge of the bed and stared at the floor.

Growth

June

DANIEL WU AND his aide, Pamela, were sitting at a table in the third grade classroom next to the display of the atmosphere of the earth. The school year was coming to a close. They reviewed their "to-do" lists and were ready to wind down the year. One more test to administer next week. Then they only had to report the final grades, take down the displays and remove their personal property. Daniel expected he would return the next year, but with budget cuts, restructuring and such, one never knew for sure. It was better to take everything that he owned or created out of the classroom just in case he had to relocate to a new job, or none at all.

Pamela, on the other hand, had no expectations to return. Yes, she would like to, but she had very little choice or guarantee of working in the same class for more than a year or even at the same school. Not in these economic times. Her volunteer time may be limited even more next year by her job demands, and the school could scale down classes and teachers considerably. She loved her class, loved her students, and loved being such an integral part of their education. She would just have to wait and see what happens with her husband's job. She knew she may have to go back to work full-time, if she could even find a job at her age.

Neither of them wanted to leave for the summer. They wished the class could continue for a few more months so they could see if Bonnie continued to excel as rapidly as she had the last three months.

"This is really quite remarkable," Daniel said. "Her scores are off the chart."

Pamela held the test papers in her hands scanning from one to the other, page after page, as though she was searching for any mistake the child may have made. "These are perfect. Every one." She laid the papers on the table. "How can she do this?"

"I don't know, Pam." Daniel scratched his head as he scanned the documents. "Three months ago she was mediocre at best. Now, she's a genius." He laid the papers out and grabbed another pile. "The other students are nowhere near her. Look at these." He handed them to Pam. "They are average across the board."

Pamela glanced at the papers. "She was an underachiever at best. I mean she isn't the most physically fit child in the class either. She has leukemia and struggles to even attend class half the time. She's missed a bunch of her assignments, yet you wouldn't know it by looking at her test scores or this." She held up a paper. "IQ of 143."

"Genius," Daniel replied.

"I don't get it, Dan. I just don't get it."

"Neither do I, Pam."

Pamela walked over to the window and stared at the kids in front of the school either waiting for their parents to pick them up or lining up at the bus stops. "How can she show that much marked improvement in her grades in such a short time?" Pam noticed Bonnie was being picked up by her mother. She watched as the little girl climbed into the car.

"Usually it's a buildup over time as concepts, theories, and ideas link together and make sense. But this ..." Daniel held up the I.Q. tests and continued. "... this is a disconnect for me. She struggled terribly for most of the year, and now she's a genius?"

Pamela turned back to Daniel. "What do we do?"

He thought for a few seconds as he gently placed the paper on the stack. "I think we should take this to the principal and let her decide where it goes from here. I don't even know if I will be here next year. At least we can make it a part of her file and have it follow her through her education."

"I'd like to see what happens to her, Dan."

"So would I, Pam. So would I."

—*—

"Well, this is quite a day for you, isn't it Mac?" Melanie Grimes smiled as she sat with the old man and discussed the plans for the day. "Jennifer will take you to your appointments and bring you back when you finish."

"Nothing like having your own chauffeur to escort you around town on such a beautiful summer day, eh?" Mac flashed a big smile at the aide and continued.

"Technically, it isn't summer yet, Mac," the aide said.

Mac turned to her. "I know that. It isn't summer until the earth is at aphelion at 10:51 A.M. Greenwich Mean Time on June twenty-first." Mac noticed both women were shocked and staring at him. "Aphelion is where the earth is at its furthest point away from the sun in its elliptical orbit," he said as he tried to demonstrate the elliptical orbit of the earth around the sun with his hands. The women continued to stare. Mac realized his explanation was futile. "Any chance we can stop by and watch a ball game? I haven't been to a good baseball game in years."

The aide glanced at the administrator who shook her head slightly. "No. I don't think we will have the time, Mac. But we could take in a milkshake at JoJo's drive in," the aide replied.

Mac smiled. "That's a date."

Melanie picked up a piece of paper and handed it to the aide. "Here's the itinerary. Mac has several doctors to visit and quite a few tests to take. They should be able to tell us why you are making such an excellent recovery, Mac."

"Sounds fine with me, Melanie. I look forward to a little mental exercise today. Do you think we can stop by the library and check out a few books? I'm getting pretty bored here and need something to keep me occupied." Mac grabbed his jacket and stood to leave.

"I think we can manage that, Mac," Melanie replied. She turned to the aide. "I just don't want you two gone all day. We have a lot of work here so I need you back."

"Yes ma'am," the girl replied.

"O.K. Then I will see you two in awhile. Have a good time."

"I plan on it!" Mac said as he left the room without his walker. He was looking dapper in his suede jacket, fedora and cane, walking with the energy and agility of a man many years younger.

"Unbelievable," Melanie whispered as she watched them leave the room.

—*—

Quinton Lemolo was sitting at his kitchen table reading when Abbey, his wife, walked in. She was seven months pregnant and showed it. "What are you reading now?" she asked as she opened the refrigerator door and took out some asparagus spears.

"Blackjack."

Abbey peered over the opened door. "Blackjack?"

"Yeah. 'The World's Greatest Blackjack Book'." Quinton held the book up for Abbey to see the cover. "It's a book on the strategy of playing blackjack."

Abbey closed the refrigerator door. "Why are you reading *that*?" She was obviously irritated.

"Because I'm tired of barely making ends meet. We are the typical paycheck-to-paycheck family, and I deserve more," he said as he laid the book on a stack of books on the table.

"*You* deserve more? Quint, what's going on with you?" Abbey walked over to the table.

"Me? Nothing. I just want to do better than working at a sewer plant for the rest of my life."

Abbey grabbed the book from his hand. "And this is better?" she said as she held it up.

"Yes. It is. It's better because I think I can make a lot of money as a professional gambler."

"A gambler? Quinton. That's not who you are." Abbey tossed the book on the stack of books.

"How do you know who I am? We've been married less than a year, and you think you know everything about me?" He stood up and walked across the room, agitated and upset.

"I know who I married and who the father of my child is, and he is not a professional gambler!" she said as she pushed the books off the table and watched them crash to the floor.

Quinton turned around to see his pregnant wife upset, and now crying. "O.K. O.K." He walked quickly to her side. "I'm sorry, Abbey." He put his arms around her and pulled her close. "I'm really sorry, babe."

Abbey sobbed and sniffed. "How could you even think of something like that, Quint? You're a good man. A *good* man. That's why I married you."

"Sometimes I feel about as dumb as a bag of hammers. Abbey, I won't do it." He hugged her close.

"You won't?"

"No. I'll stay at the plant and look into another career if I need to, but not that," he assured her.

"I just can't see you doing that, Quint," she said. "Not with our baby on the way. We need something dependable. Something we can count on, Quint. Not this," she said as she pointed to the pile of books on the floor.

"I know. We will." He took a deep breath. "It's not a bad thing, Abbey … being a professional gambler. It doesn't make a person bad, you know."

"I know, but it's not you." She looked up at him and through her tears and said, "Do you promise you won't do it?"

Quinton smiled and wiped a tear off her cheek. "I promise." He pulled her close and looked at the books lying on the floor behind her, calling to him.

—*—

The bedroom had two chairs turned over on their side next to the bed. A large blanket had been pulled off the bed and draped over the chairs. A light could be seen moving under the blanket.

"They'll never find us here," whispered a young boy. He was six years old and the leader of the dynamic duo. "If we stay real quiet, we can plan our next invasion of the castle and they will never know. We'll surprise them."

His younger accomplice was wide-eyed and eager to participate. "An we can get some gold from the safe ... an then give it to the poor peoples who don't have none ... an ..."

"Alright." Alice Kruger was standing at the door leaning on a crutch. "And just what are you two hooligans planning?" She tried to look and sound stern, but her smile surfaced as the older boy peeked out from under the blanket.

Alice was in her forties with a big smile, red hair, bright green eyes, and a distorted body. She suffered from a genetic disease called Becker Muscular Dystrophy. The symptoms are deterioration of the muscles of the legs and pelvis area, resulting in twisted limbs. Walking was an effort. Alice often used two crutches that wrapped around her wrists. She would twist and turn as she struggled to move one leg in front of the other. She was able to maneuver about the house or small space with one crutch and perhaps a table, chair, or someone's arm. There is no known cure for BMD. Treatment is aimed at control of symptoms to maximize the quality of life.

"Nothing, Auntie Alice." The older boy, Leonard, poked his head out from under the blanket. "Honest."

Alice was the Director of Data Research and Technology Department for the Centers of Disease Control in Atlanta, Georgia. She received her Masters in Data Management from the University of Southern California and joined the CDC where her career excelled. She was an integral part of the CDC's "19Q" team that battled the genetic mutation of the *PNKP* gene that resulted in nearly 60,000 deaths a few years ago.

The younger boy of five years old peeked out next to his friend. "Yeah, mommy. We aren't going to rob the safe or nothin."

Alice chuckled as she watched the boys crawl out from their fort. She loved children, especially her son, Joey. She always wanted to have children, but didn't marry until late in life when she met Carl Kruger several years ago while working together on the "19Q" team. They agreed that at her age and having BMD it was too risky to have a baby; for her and the child. So they adopted.

"I sure hope not," she said. "It isn't right to steal, Joey."

"We weren't stealin, mommy."

"We weren't, Auntie." Leonard looked down and continued. "Just pretendin."

Alice smiled and squatted down to eye level with the boys. "And what were you pretending?"

Joey replied excitedly. "This is our fort and that is the castle an we were goin to take the gold from the evil king and give it to the poor an…"

"Whoa, whoa. Hold on." Alice reached out and took her son's hands. "That's a pretty big job there." She pulled him into her arms and reached around her nephew. "I think you two are heroes, and you should have something to eat to stay strong and keep doing good deeds."

"Can we have some cookies?" Leonard asked.

"No. Not cookies. But you can go to the kitchen and see what looks good. I'll be right there." Alice gently nudged the two boys toward the kitchen as she rose and started down the hall toward the bedroom.

The boys raced down the hall and into the kitchen where Doctor Carl Kruger was seated at the table in the breakfast nook reading the paper. He was a man among men. At six foot four and 260 pounds, he was a big, African American man. He played college football as a linebacker and was built like a rock with broad shoulders and big hands. He was in his early fifties and had a shaved head and a Van Dyke beard. Carl thought of shaving off the beard and growing his hair back, but Alice vigorously objected. She fell in love with Carl the moment she laid eyes on him. But it was his voice that melted her. "I hope your feet are clean since mom just mopped this floor." His deep voice boomed through the kitchen and startled the boys. He didn't look up as they stopped to look at the bottoms of their feet.

"We don't have no shoes on, dad."

Carl put the paper down and corrected his son. "You mean you don't have *any* shoes on, Joey."

"I know. That's what I said." He turned and opened the refrigerator door as he and his cousin peeked in while Carl returned to his paper with a chuckle.

"Dad?"

"Yes."

"Can we have some yogurt?" Joey knew he had to ask permission to take things out of the refrigerator. Mom told him he had to watch his snacks during the day. He also knew that his dad was the official aficionado for Crystal Creamy Yogurt. It was his favorite dessert, and he prized it as highly as others might a bottle of fine wine.

"I don't know, son. That's pretty good stuff, ya know." Joey could see the smile on his dad's face as he lowered the paper. "Sure, go ahead. Just save a little for me, please."

Carl once again returned to his paper as Leonard anxiously reached into the refrigerator, grabbed the yogurt, and quickly pulled it toward him. The large container was too big for his little hands and slipped out of his grasp, crashing to the floor.

SPLAT!!

Carl lowered the paper to see his nephew with a look of terror on his face as he stood over the pile of yogurt. The carton had burst open and pink, strawberry yogurt splattered at least three feet in every direction, including on his feet.

"I'm sorry, Uncle Carl." The poor lad was almost ready to cry as Carl slowly stood up and approached the scene. The tall man cautiously surveyed the damage. His valuable yogurt was all over the tile floor, Leonard's feet, and a little on the rug by the sink. He looked at Leonard and assessed the best way to approach the catastrophe. Joey watched intently wondering what his father would do. "I didn't do it, dad."

"I know, son."

His six foot four frame towered over the boys like a giant as he put his large hands on his hips. His muscular build stretched the sleeves of his shirt giving him an ominous appearance to the children. The boys looked up at him fearful that he was ready to pounce. Carl carefully crouched down so he was at eye level with the boys. He looked them in the eyes and saw tears starting to form. He gently placed his large hands on each boys' shoulder.

"Well, this sure is a big mess."

"Uh huh." Leonard was ready to burst into a stream of tears as he looked down at the yogurt splattered on the floor.

Carl shook his head as he said, "I guess there's only one thing we can do."

Leonard looked up at his uncle with his big, brown eyes, wondering what type of punishment would befall him. Would his uncle spank him, or ground him, or ask him to leave or …? Carl knelt down onto his hands and knees …… and took a giant lick of yogurt off the floor. He looked up at the boys with the pink yogurt smeared across his black face, beard and chin. Leonard was aghast in disbelief. He had a look of astonishment on his face.

"Woof." Carl's deep voice boomed through the kitchen.

A huge smile stretched across Leonard's face as he knelt down next to Carl and started licking the yogurt off the floor. Within seconds, Joey joined in. The three of them knelt on the floor and licked up yogurt like dogs, except they were giggling, laughing and occasionally barking.

Then, Alice walked in. "What are you doing?"

Joey was laughing so hard, Alice could hardly make out what he was saying. "Eating yogurt, mommy."

"Are you crazy?" she asked.

Carl looked at the boys and barked. Then, he reached over to the drawer and pulled out a spoon and said, "You're right, honey. These will work a lot better."

With that, both boys burst out laughing as they watched the big man eat yogurt off the floor with a spoon. Alice reached into the drawer with disgust, grabbed two spoons and handed them to the boys. "At least you can stop barking."

Alice was glad she had just mopped the floor earlier that day and wasn't upset that she would have to clean it again. She laughed as she watched Carl and the two boys eat yogurt off the floor. She relished the moment.

—*—

The office was almost empty. There was one small window looking out to a bush that blocked the view. A large bookshelf stood as a testimony to the doctor's education. In the middle of the room sat a square table with two chairs sitting opposite each other. Stanley MacKenna Thorne sat in one chair with a pencil and notepad in front of him. He was relaxed and listening to the instructions given by Doctor Eldon Northridge, seated across from him. "I'm going to show you a series of letters for three seconds. You will need to write them down the best you can. It's a simple test. O.K.?"

Mac sat forward in his chair. "Sure, Doc."

"Ready?"

Mac picked up the pencil. "Shoot."

The psychologist turned over a card. The letters 'UM' were on the card. He left it visible for three seconds and turned it back to the blank side.

Mac didn't move. "You can write the letters down now, Mr. Thorne."

"Please, call me Mac."

"OK. Mac, you can write them down now."

"Do I have to?"

"No, we can go to the next card. We need to go through all of them, so just do the best you can." Doctor Northridge could see that Mac was struggling to remember the letters like many of his patients do. He could recall just a handful of patients who could not even get the first series of letters. Mac didn't even try. It looked like he was going to be another failed patient. "Here's the next card." The doctor turned the card over to reveal the letters 'TZLD.' After three seconds, he turned the card back.

Mac stared at the table, tapping the pencil on the pad.

The doctor turned over the next card, KXCEJO, waited three seconds, and turned it back. Mac watched. Then the fourth, fifth, and sixth cards. The doctor patiently turned each card over, waited, and returned the card to the blank side. Each time he waited for Mac to move, to write something down, to try.

Nothing. Mac never moved. He just tapped the pencil on the pad and watched. "O.K. I think we are done ......"

Mac leaned forward and started writing on the pad feverishly.

UM

TZLD

KXCEJO

AVCYISEH

LBFQRPMAUX

ZQECTBUMONRV

Doo-Dah

Mac leaned back in his chair and tapped the pencil on the pad, and smiled.

The doctor realized Mac was tapping the pencil to the song, "Camptown Races."

Beginnings

September

QUINTON DIDN'T KNOW why he had such a good memory. All he knew was that he could remember everything. *Everything.* He flipped the edge of the cards toward him and said, "Hit me."

The dealer turned over a card. "Queen." Quinton glanced at her and thought the middle-aged woman looked like she could have been Native American. It would seem appropriate since it was an Indian casino.

"I'll stay." Quinton laid the cards face down and sat back. He remembered every card played out of this deck so far. Four players plus the dealer, three rounds played, twenty four cards out, the fourth round in front of him. The odds of the dealer having a face-card down were very high. And, with a five showing, it looked like Quinton was going to win … again.

The woman flipped her bottom card over. "Dealer has fourteen." She tapped the table twice and turned a card over. "Nine. Twenty-three. Dealer bust." She reached out, flipped over each player's cards one at a time and stacked chips to match their bet. Quinton had quite a large stack on this bet. "Nice bet," she said.

Quinton raked the chips over to his ever growing stack. "Thanks."

Quinton felt bad that he had lied to Abbey. He told her he was going to a class to learn about some new software the plant installed for managing the new tertiary filters. He was partially right. He learned about the software, but not at a class. Instead, he told his supervisor that he had to take care of their baby, Jenny, and couldn't make the class. So, he asked for the materials. He was able to read them last night and knew everything he needed to know. He didn't need the class now. He needed to keep winning.

A lie on a lie.

And win he did. Which made his 'feeling bad' diminish rather quickly. He was more than a thousand dollars ahead. And it took less than an hour.

The pit boss walked over, stood behind the dealer and watched Quinton stack his chips. Quinton could see him out of the corner of his eye and knew he was being watched. The books he read about gambling said he needed to incorporate 'levels of loss' to avoid detection. Different authors had different theories about how much to lose and when. Quinton sorted through the different options and placed a nice bet as the dealer shuffled the cards. "I'm ready to break the bank," he said. "Baby's gonna get new shoes."

The dealer dealt the next hand as the pit boss watched. Quinton barely looked at the cards. The dealer had a six face up as the top card. When the dealer came to Quinton he said, "Hit me." The dealer turned over a six. "Again." The dealer turned over a queen. "Busted." Quinton tossed the cards face down and grabbed his chips. "Well, I better leave while I'm ahead." He tossed a fifty dollar chip to the dealer who took it, tapped it twice on the table and placed it in a slot. "Thank you, sir," the woman said. "Good luck."

The pit boss turned and walked to the next table.

The dealer turned Quinton's two bottom cards over to place them on the bottom of the discard pile. They were a seven and three. The dealer watched as Quinton walked over to the cashier's window with his winnings. "Smart kid," she mumbled.

—*—

Golden Pond Memory Care Facility was abuzz. Today, one of their residents was going to be discharged home. It was a rare event indeed. Usually, when someone is checked into the facility, they stay until they either die, go to the hospital with a medical condition where they die, or are taken home by a loved one to die. Either way, death was usually the only way out.

Until today.

"Hi, Mac." Melanie was all smiles. "This is quite a day for you, isn't it?" she asked.

"Ma'am, it most certainly is," Mac said as he stood and took off his fedora in a sweeping motion and bowed low.

"Well. I can see you are pretty excited about this," she said.

"Of course I am. I'm going to move in with my grandson and start living again." Mac put his hat back on with a tap of the hand and ran his fingers around the brim. "I feel young again," he said with a broad smile.

"And you look young, old fella." The voice came from around the corner. A middle-aged man peered around the corner and smiled. "You ready, Grandpa Mac?"

"Anthony!" Mac reached out and grabbed the man's hand and shook it heartily. "More ready than you will ever know." He pulled him close and gave him a strong hug.

"Well. Looks like we better get you two on the road." Melanie reached for several papers stacked on her desk. "I need to have Anthony sign these documents for the discharge since he is your legal guardian."

Mac became irritated. "I don't need a guardian. I can take care of myself."

Anthony placed his hand on Mac's shoulder. "We know you can, grandpa, but until we get the guardianship terminated, I still need to sign the papers. It's no big deal."

"It is to me." Mac watched as Anthony scanned the papers and signed at the bottom. "O.K., boy. Let's go." Mac grabbed his suitcase and cane and started toward the door.

"Here. Let me get that for you." Anthony reached out to take the suitcase away from the old man, but he wouldn't let go.

"I've got it, Tony." Mac pulled it back.

"But Grandpa, I want ..."

"Listen, son." Mac stood square to the young man. "Part of being a man is doing things your own way to the end. It's counting on no one but yourself and God. That's what got me through the war, through Mabel's death and through this place. Now, I'm going to carry my own bag out this door and to the car. From there, you can have it." Mac grabbed his suitcase and started out the door. He stopped and turned back to Melanie, placed the suitcase on the floor and walked over to her. He took his hat off and then took her hand. "I want to tell you how much I appreciate your kindness, gentleness and sincere appreciation for us old folks. You are an angel." He leaned down and kissed her hand. "I will not soon forget what you have done for me." Then he turned, grabbed his suitcase and started for the door before Melanie could say anything.

"Bye, Mac," she whispered as he headed out the door.

Melanie could hear the cheers from staff and other residents as Mac walked between the throngs of people lining the sides of the hallway to the exit.

—*—

Doctor Carl Kruger loved his life. He was happily married to a wonderful woman who adored him, and he her. They had a son they adopted who was growing into a fine little boy. His job was demanding yet extremely rewarding. He was in great health, financially secure and had a good reputation among his peers. Life was good. So good.

Carl smiled as he pulled up to the large, modern brick building spattered with windows and a flag flying on top. It was several stories high and a block long. It was the Johns Hopkins Bayview Medical Center in Baltimore, Maryland. The building housed the Memory and Alzheimer's Treatment Center, one of three clinics under the supervision of the JHU Alzheimer's Disease Research Center.

Carl worked for JHU for years. About four years ago he worked for the JHU Institute for Cell Engineering, or ICE. It was a relatively new concept in medicine that supported and housed scientists working to understand how cells' fates were determined and to harness that information in order to select, modify and reprogram human cells; genetic engineering and analysis. Dr. Kruger was working to identify multiple mutations in the *PNKP* gene that resulted in severe neurological diseases, when he was called to the Centers for Disease Control in Atlanta to work with a team of scientists to combat a new disease that was rapidly killing people. That was where he met Alice.

The experience he garnered working on the "19Q" team, as it was called, enabled Carl to excel. He was recognized for his significant contributions to the team's success, and became renowned for his understanding and research of biomedical genetic engineering. That was why he was promoted to his new position two years ago; Project Coordinator for the Alzheimer's Genetic Research Team. In short, he was given a team of doctors and scientists to specifically research genetic influences of Alzheimer's patients to determine if a genetic cure could be found. It fit perfectly with his prior research of the *PNKP* gene. The only difference now was that he focused on the *APOE* gene of the same chromosome; chromosome nineteen.

The woman at reception smiled as Carl stepped into the lobby. "Good afternoon, Mr. Kruger."

Carl quickly glanced at her name badge hoping she wouldn't notice. "Hello, Anna. How are you today?"

"Very well, sir. Thank you."

"That's great," Carl said as he walked past her station to the elevator. The lady admired the muscular man in a tailored suit. It fit him well.

As Carl rode the elevator, he thought of the research this center was doing to treat thousands of people with dementia, Alzheimer's, brain injuries and brain vascular disease. Carl was pleased that he was an integral part of helping thousands of people live better lives. He was hopeful that their collaborative efforts with genetics, neurology, psychiatry, and geriatrics could someday lead to a cure for Alzheimer's and other dementia related diseases.

Carl stepped out of the elevator and was greeted by another receptionist. "Mr. Kruger. Good to see you."

"Thank you, Jane. Is Casey in?"

"He is."

Carl nodded as he walked past the receptionist toward the facility administrator's office. He was anxious to discuss the team's latest findings and enroll Doctor Casey Knowles with supplying patient information from his facility pertinent to their current research; tangle and synaptic deterioration rates. Carl didn't particularly like meetings like this. The administrator was often busy with mundane tasks and focused on their narrow world of the facility. Carl was sure this meeting would be an interruption to Casey.

That was why Carl passed on the offer of becoming a facility Administrator and chose to be a project leader in the Alzheimer's Disease Research Center instead. He didn't want to manage a facility like Casey. He was a "hands on" type of guy. He liked getting his hands dirty and relished the opportunity to continue to learn, grow and apply his ever-advancing knowledge. Carl felt like a beggar on the corner each time he met with the administrator of any facility asking for something. This time, however, he hoped Casey would see the mutual benefit of sharing patient information and fully participate.

Carl stepped through the door and closed it behind him. "Hello, Casey."

—*—

It was a beautiful Saturday afternoon. September is a fabulous time of year in the San Joaquin Valley of Central California. It's the time of year when people are outside doing chores, riding bikes, going on hikes, boating at the lake; most everything except sitting in their office laboring over a stack of papers like Doctor Eldon Northridge.

Mac's discharge from Golden Pond Memory Care Facility was unique. He was going home to live, not die. Eldon had never seen anyone recover from Alzheimer's Disease like that. Never. It was unheard of, until Mac. He was the first. He showed significant signs of memory improvement to the point of being discharged from the facility.

"This is remarkable," Eldon mumbled as he sorted through the documents examining the test scores and chart notes.

"It is, indeed," his wife, Stephanie, agreed. She, too, had a Master's Degree in Psychology and was a partner in the business. Her forte was child psychology. Her cases were mostly comprised of dysfunctional children who suffered from varieties of traumas; abuse, neglect, brain injuries and mental illness. Occasionally, the doctors would share stories of patients they had seen, or treatments they administered. Patient confidentiality was absolute outside the office. Inside, it was pillow talk. "This is unheard of, Eldon. No one recovers from Alzheimer's or any naturally occurring dementia like this."

"I know." Eldon shuffled through the documents over and over again, reading aloud as he went. "Stanley Thorne; mini mental[3] went from a nine to twenty-six." He stopped and held the document up. "Nine to twenty-six. That's impossible!"

Stephanie picked up the paper and looked at it. "I know. But here it is. Twice."

"He was a near vegetable barely functional, and now he is near normal."

Stephanie glanced through the rest of the file. "The other tests and chart notes indicate steady improvement in cognitive abilities and functions from April on. Everything is showing marked improvement. Behaviors are normal; mobility has improved; minimal to no assists with ADLs[4]. What is going on, Eldon?"

He took off his glasses and rubbed his face. "I don't know. I have no idea." He walked over to the large window overlooking the valley. Large, billowy clouds floated across the valley toward the Sierra Nevada Mountains. The brown, dry grass covered the hillsides spotted with valley oaks whose leaves were just starting to turn, creating a canvas of colors. Eldon never noticed. He was deep in thought. "This is huge, Stephanie. Huge. Something is happening here that I just can't put a finger on. I suspected it when I tested Thorne three months ago." He turned back to his wife. "I thought it was an aberration; a fluke. But this ..." He walked back to the table and grabbed the papers. "... this is saying it isn't. It's ... it's ..."

"Unsettling," she replied. Eldon looked at her as she continued. "Almost scary."

---

3 The Mini-Mental State Examination (MMSE) or Folstein test is a brief 30 point questionnaire test that is used to determine levels of cognitive impairment and capacity. It is commonly used in medicine to screen for dementia. Scores below 25 indicate levels of cognitive impairment.

4 Activities of Daily Living—A term used in healthcare to refer to daily self-care activities within an individual's place of residence, in outdoor environments, or both. Ability or inability to perform ADLs is often used as a measurement of the functional status of a person.

Eldon tossed the papers on the table. "We need to run some more tests to find out what's going on here. I think I should have a PET[5] scan run."

"I agree."

Eldon picked up one of the tests and looked at it again. "How is this happening?"

—*—

The yellow school bus pulled up to the visitor area of the Applegate Wetlands. The door opened and Mr. Daniel Wu and his aide, Pamela, stepped out onto the gravel parking lot. They were excited to have another year together at the same school.

"O.K., kids," Pam said as she clapped her hands. "I need everyone to line up here in rows of twos." She motioned for the lines to form parallel to the bus.

Daniel Wu flipped through some papers and handed one page to each child as they lined up. "This is the map of the Applegate Wetlands. Don't lose it." When the children finished fussing about who was going to stand next to whom, Mr. Wu continued. "This field trip is your introduction to the third grade. Does anyone know what 'introduction' means?"

A young boy in the front of the line raised his hand. "Yes, Nathan."

"It means meeting something for the first time."

"Correct. Thank you, Nathan." He held up the map. "Every third grade class starts with a field trip to the wetlands. Today, you are meeting the Applegate Wetlands. This will be the first of several field trips that this class will take to the wetlands." Several children cheered as the teacher continued. "A few rules. Do not throw anything into the water or off the walkway. Animals live here."

---

5 Positron Emission Tomography—Medical procedure that measures emissions from radioactively labeled metabolically active chemicals that have been injected into the bloodstream. Used to map neurotransmiter activity and diagnosis of neuron-damaging diseases such as Alzheimer's.

Pam chimed in. "That's right. You wouldn't like it if someone came by your house and threw their garbage out, would you?"

"Unless you lived at the dump," one boy mumbled. A couple of kids nearby giggled.

"What was that?" Pam asked. No one answered.

Daniel continued. "No horseplay here. Stay on the walkway at all times. If you break any of these rules, you will be brought back to the bus to sit while we finish the tour." The children quieted down and paid attention, because no one wanted to come back and sit in the bus alone. "We will be viewing a variety of wildlife and plant life and learning how they depend on each other to survive. O.K., are there any questions?"

No questions.

"Follow me, and stay together in pairs." Daniel Wu led the procession to a gate bordered by two garbage cans with a sign that read,

Leave your garbage in the can ...
We don't want it in our home ...
Signed—The wildlife

He paused as Ms. Engle's fourth grade class was leaving the wetlands tour. "Hello, *Ms.* Engle," Daniel said, emphasizing the 'mizzz' for the class.

Nancy nodded and returned the salutation. "And a good day to you, *Mr.* Wu." She flashed a smile knowing the teacher wanted to demonstrate respect to teachers and elders. "Enjoy your tour, sir." She walked past the third grade class leading her group of fourth graders.

Daniel Wu watched as the children filed past in pairs. The last pair in the line was two girls; one was Bonnie Howe. She looked directly at Daniel as she passed, and slightly smiled.

Daniel felt a shiver down his spine.

Scans

November

"I'M HERE FOR a scan."

The young lady behind the counter glanced at the elderly man and asked, "Name, please?"

"Stanley MacKenna Thorne. People call me Mac." He flashed a big smile.

"Date of birth?"

"March sixth, nineteen twenty-three," he said proudly.

The young lady looked up. "You're ninety-one?" She was shocked. "You look much younger, Mr. Thorne."

"Mac, please, call me Mac. Thank you for the compliment."

"Okay. Mac. Your doctor is Northridge?"

"Yes ma'am."

"O.K. If you could have a seat right here as I get your information together."

Mac sat down, leaned back and folded his hands in his lap. He looked like he had called a meeting. He exuded confidence and command presence.

The young lady handed Mac a clipboard. "This has your primary information on it." Mac scanned the board and made notes as she spoke. "I'll need your insurance card." He handed her his card and she continued. "I'll need twenty dollars for the co-pay." Mac handed her the money. "I'll be right back with your receipt."

Mac continued to scan the board and make notes. The young lady returned with a paper in her hand. "Here's your receipt. If you could please sit over there and check your date of birth, address, contact information, medical history, and other areas indicated in yellow. Please review it and make changes as appropriate. When you are finished you can ………"

"I'm done." Mac held the clipboard out waiting for her to take it.

"Done? I'm not sure you understood what I was …"

"Young lady." Mac leaned forward in his chair and pushed the clipboard toward her. He had a stern look on his face and spoke slowly. "I … said … I … was … done." He smiled as she took the clipboard from him.

The girl glanced over the paper. "Uh … yes, you are. That was fast. I don't think I've ever …"

Mac cut her off. "How long will it be until I see the doctor?"

"Uh. Just a few minutes, Mr. Thorne. He will be right with you."

Mac leaned forward and smiled, again. "Please, call me Mac." His smile turned to another stern look. The young lady watched as he walked across the room, changed the channel on the television to market reports and took a seat. Mac was watching the TV when a side door to the back office opened and a little girl walked out with her mother.

Bonnie Howe was done with her tests and going home.

—*—

Quinton rummaged through the bathroom drawer. He was rapidly becoming agitated, as was often the case. He knew the pills should be in this drawer, left side, three rows back from the front. "Where's the migraine medicine, Abbey?" he yelled as he continued to push bottles around the drawer.

No answer.

Quinton slammed the drawer and held the right side of his head. "Abbey!" He turned and started down the hallway. He could hear the baby crying as he approached the nursery. "Abbey!" Quinton rounded the corner and saw Abbey holding and rocking the crying baby. "Can't you hear me?"

"No. Not with Jenny crying in my ear." Abbey slowly rocked her infant daughter. "I think she has colic."

Quinton reached out and rubbed the baby's back and leaned toward her. "It's O.K., Jenny," he tenderly whispered in her ear.

The baby stopped crying and settled down after she heard her dad. Abbey smiled. "She loves your voice."

"And I would love to find that migraine medicine. Where is it?"

"In my purse. I was going to get another bottle and took this with me to make sure I would get the right thing." Abbey handed the baby to Quinton and walked out of the room. Quinton whispered to his daughter, "Did I ever tell you the story of the princess that lived a long time ago in a land far away and had a huge tree in …"

Abbey walked back in and interrupted him. "Here they are." She handed him the bottle as she took the baby back into her arms.

"Thanks." Quinton opened the bottle and took two pills without water, throwing his head back to swallow them.

"Quint. Are you getting another migraine?"

"No. I thought these were sour balls. Of course I am!"

"Honey, that's the fourth one in two weeks. I'm worried." Abbey's concern clearly showed. "I think you should go see the doctor and have it checked out."

"Normally, I would put up a fight and not go, but this is really beginning to bother me." He sat on the side of the bed. "It's a burr under my saddle and I'm getting tired of it. I'll call tomorrow and set up a time."

"No. You'll call today. Right now."

"O.K. I'll call now." He stood and walked out of the room.

Abbey held her daughter and whispered to her. "Something's going on with your daddy. I hope he's alright." She rubbed the baby's back as she spoke. Suddenly, the baby let out a huge burp. Abbey was startled by the sound, then giggled. "That's better."

—*—

Mac was bored. He just left the prep room after waiting thirty minutes for the nuclear medicine to work its way into his system. He felt fine, except being bored. *They could have provided some good books, or movies or puzzles or something other than those decrepit magazines,* he thought. Now he was going to the examination room down the hall.

The examination room was large and dreary in color. A cylindrical tube about six feet long was in the middle of the room. A table with a white sheet on top of a pad protruded from the end. A large arm connected the tube to a track on the ceiling.

"If you will please sit here for just a moment while I get the technician. He will brief you on the procedure." The young girl directed Mac to a chair sitting by the door at the end of the tube.

Mac reached behind him and held his gown together as he sat. "A little breezy here," he said as he smiled.

A middle-aged man rounded the corner with a pad in his hand. "Mr. MacKenna?"

"That's me. Call me Mac, please."

"O.K., Mac. I'm Larry." They shook hands. "Sorry I was a little late. I had to make sure my last patient's scans were accurate." Mac didn't say anything and the technician continued. "We are going to conduct a PET scan on your brain. Did the doctor tell you about this?"

"He did."

Larry surveyed his patient and thought it would be best to re-explain the details of the procedure. "This is a simple procedure with complex results," he began. "First, we injected you with radionuclide about thirty minutes ago. Do you remember?"

"Of course I do." Mac was perturbed that the technician thought he was unable to understand what was happening. "The radionuclide will concentrate in the brain where the Positron Emission Tomography scan will measure the metabolic activity of the brain by measuring the gamma rays emitted from the positrons as the radionuclide breaks down." Mac smiled. "How am I doing, Doc?"

The man was stunned. "Uh. Fine." He was searching for a response. "Are you a doctor?"

Mac laughed. "No. Not hardly. Just an interested person." Mac leaned back in his chair, pleased that he could baffle another educated professional. "Are *you* a doctor?"

Larry was slightly offended by the way the question was asked, but he maintained his professionalism. "No. I'm a technician trained specifically in the field of nuclear imaging."

"Good. That would be helpful in this room with that machine," Mac said as he pointed to the cylinder and laughed.

"It is." Larry quickly changed the topic. "Now, the scan will take about thirty minutes, so you can just lay back and relax." He wrote a few notes on the board. "Any other questions?"

"Nope."

"Good. Then, let's begin."

Larry laid the clipboard on the table and watched as Mac walked over to the bed and lay down with no effort. "Let's go, *Doc*," Mac said, snapping his fingers. "I'm a busy man."

—*—

Mac had his PET scan just a week ago. It happened on the same day that Bonnie Howe had her PET scan. The technician administering the scans of both patients reviewed the results for accuracy, clarity and content before forwarding them to the respective doctors as he always did. He worked in this field for twenty years and administered hundreds of CRI, MRI, SPECT and PET scans. He personally reviewed the results of all of them. He understood what many of the scans indicated. His education and experience with the operation of the various modalities of nuclear medicine enabled him to interpret the results as well as a specialist. He was a specialist in his own rights. He knew what the scans were saying. And they troubled him deeply.

Both scans showed brain activity that was off the charts.

Larry knew that the neurologists would ask if the scans were accurate. That was why he drafted a summary statement with each one as he forwarded the results to the neurologists, assuring them that the tests were conducted properly and verified for accuracy.

There were three neurologists in Elk Grove who worked at Mid-Valley Neurology Center near the Consumnes hospital. They were peers and friends. Often, they would cover each other's cases if one was on vacation.

Special cases with special circumstances called for special meetings. Nothing was more special than the results of these two scans.

"I've never seen anything like this, Adam." Alexandria Harlow was the 'queen' of neurology. She was the founding partner and had the most experience in the field. Her credentials read like the alphabet. Her peers tolerated her; her friends adored her. Alex, as she liked to be called, enjoyed a challenge.

Adam Daniels agreed. "This is unbelievable," he said as he waved the picture of the brain. Adam studied neurology at Johns Hopkins under some of the greatest in the field. His youth and unconventional approach labeled him a maverick when, in reality, he was a genius. "This is from a ninety-one year-old man who had been diagnosed with Alzheimer's?"

"It is," Alex replied. "Amazing, isn't it?"

Mac's scan revealed a brain functioning very effectively for a ninety-one year-old man. Very effectively. It looked more like a teenager's brain than an elderly man with any dementia disease, much less Alzheimer's. The red areas were abnormally large during all functions and stimuli. There was no indication of any abnormal or deficient neural activity. None.

"Yes. Remarkable," Adam said. "Alzheimer's slowly destroys the synaptic connections in the brain and causes atrophy. How can he have this level of neurotransmitter activity if he has Alzheimer's?"

"Impossible, I know," Alex replied. "We know that there is confirmed neurogenesis[6] in several areas of the brain, but nothing on this scale or this definitive. Brain cells die and mostly stay dead."

Adam laughed. "Mostly stay dead? You sound like the old man on the Princess Bride." Adam laughed again. "Mostly stay dead."

"Hey, I'm a neurologist, not an English teacher." Alex laughed with him. "A mostly live one."

"Sounds like a party in here." A short, middle-aged man walked over to the table to see the pictures without invitation. He didn't need one. He was the third neurologist at the facility; Doctor Howard Huffman. "What's this?" he asked as he held up the pictures of Mac's scan.

---

6 Neurogenesis –The process by which neurons (electrically excitable cells that process and transmit information by electrical and chemical signaling) are generated from neural stem.

"That, my dear Huffy, is the scan of a ninety-one year-old man with advanced Alzheimer's." Alex knew she dropped a bombshell, and waited for his response.

"Bull. What is it, really?" Howard asked again, this time looking to Adam.

Adam smiled. "It is as she says, dear sir," he said as he pointed to Alex who, again was smiling.

"You guys are trying to pull some kind of joke on me, aren't you?" Howard said as he tossed the pictures back on the table.

"Howard." Adam knew if he used Howard's proper first name, he would garner his attention, unlike Alex who liked to toy with him. "They are of an elderly man who was diagnosed with Alzheimer's. Hard to believe, isn't it?"

"Impossible," came the reply.

Adam pushed the picture aside and pulled one out of a stack next to him. "Here's another one just as baffling." Adam held up one of the PET scans from Bonnie Howe's scan.

Alex took it and quickly lost her smile. "There has to be a mistake here." Howard leaned over her shoulder to see the scan.

Adam became very serious. "I thought so too, but the technician did everything by the book, as he always does, and double checked it. I told him it must be a fluke. But, here it is."

Bonnie's scan indicated a brain functioning at near capacity.

"I don't think this is even possible," Howard said.

"I know. Her parents said the girl is an absolute genius. She wasn't always like that, from what I am gathering." Adam said. "They have had her through a battery of tests and finally have come to the conclusion something has changed in the last year. She went from a mediocre intellect to genius. She has a photographic memory, and I mean photographic. She can read a book one time and recite the entire thing verbatim, a week later."

Howard stood upright. "Did you say a recently developed photographic memory?"

"I did. Why?"

Howard took the scan and laid it next to Mac's scan. "Are these patients having any migraines or other discomfort?"

Alex looked puzzled. "No. Not mine."

"Or mine. Why?" Adam asked.

"I had a doctor refer a patient for a neurological eval because he was experiencing migraines."

"So?" Adam said.

"The doctor said the patient had recently developed a photographic memory and they are concerned it could be the cause of the migraines." Howard sat next to the two doctors. "He said the patient had a remarkable memory like none he has ever seen or heard."

"Have you seen him yet?" Alex asked.

"No. Our appointment is the first week of December."

"You should consider running a PET scan … if you believe it is in his best interest as a patient to determine the cause of the migraines." Adam smiled.

Howard laughed. "Of course it would be in his best interest to run one. Without a doubt."

Alex picked up Mac's scans. She slowly flipped through them, examining every detail as she spoke. "This is absolutely astonishing. We need to know why this is happening."

Presents

December

JUST FIVE DAYS until Christmas.

Carl was dressed and ready to go to church. As usual, he was waiting for Alice to finish up. Alice took more time to get ready because her disease made even some of the easiest tasks, like getting dressed, difficult. Carl sat in his chair to catch a glimpse of the Sunday morning football game while he waited for his wife and son. He turned the television on and leaned back.

"Carl? Can you get some cereal for Joey and fry a quick egg for me before we go?"

Carl knew he didn't have a choice. The game would just have to wait. "Sure, hon." He walked across the room and into the kitchen watching the TV all the way. Every few seconds he would peek around the corner to see the game.

Joey rounded the corner and dashed into the kitchen. "Daddy, where's the clue?" The boy scanned the counter where Carl normally left the clue every morning. Carl started a Christmas tradition for Joey that his dad did for him. Each morning on the twelve days before Christmas, Carl would write a simple clue for Joey to read. The clue would direct Joey to a place where Carl hid a small present. Joey was excited to get his morning clue from his dad.

"Did you wash your face and hands?" Carl asked.

"Yes, sir. And I washed my teeth, too," the young boy replied proudly.

Carl laughed. "Good. Nothing like a good tooth washing to start the day." He placed a bowl on the counter and filled it with cereal and milk. "Here ya go. Some good 'ol Cocoa Puffs."

"Mmmmm." Joey smacked his lips as his dad sat the bowl on the counter. "Where's the clue?"

"Right here." Carl pulled the clue out of his shirt pocket and handed it to Joey. He anxiously opened the folded piece of paper and read it.

"Rub a dub dub." He let out a yell. "I know where it is! Three men in a tub!" Joey ran down the hall and into the bathroom. Alice yelled, "Not here. Get out! And close the door!" Joey turned and ran toward the other bathroom unphased by his recent encounter. The present was much more important.

He ran into the other bathroom and found a small, wrapped package in the tub. He excitedly opened it to find a Hot Wheels truck. "Cool!" He ran back into the kitchen and hugged his dad's waist. "Thank you, dad. This is great!"

Carl leaned down and hugged his son. "You are very welcome." Joey stared at the toy as Carl continued to speak. "Son, these daily presents just lead up to Christmas morning where we get the greatest gift of all; the birth of Jesus."

Joey looked up and smiled. "I know, dad. He was born in a manger with donkeys and stuff."

Carl laughed. "Yes, there were donkeys I'm sure." Carl lifted the boy onto the seat at the bar. "Here's your breakfast. Better eat up so we aren't late."

"O.K. I'm ready." Alice said as she entered the kitchen leaning on both crutches. She had her messenger bag around her shoulder. "Where's my egg?"

"Right here, babe." Carl slid the egg out of the pan and onto a piece of toast.

"O.K., Joey. Finish up. I don't want to be late for church," Alice said as she took a bite of the egg sandwich and grabbed her bible and purse and pushed them into her messenger bag.

Joey picked up his bowl and commenced to pour the milk down the front of his shirt instead of in his mouth. "Ahhhh," he said as he wiped his

THE MEMORY PROJECT 65

mouth with his sleeve, then noticed the milk that spilled onto the floor. "Oops!"

"Joey!" Alice put her egg down and reached for a towel. "Here." She handed it to Carl.

Carl wiped the boy off and then wiped the milk up off the floor. "You know, we can clean this up later. Let's go so we aren't late."

"You're right. Come on, boys. Into the car."

Joey ran out the door and jumped into the waiting car. Alice hobbled out on her crutches with her bag and coat. Carl walked with her and opened the car door for her. "Ma'am."

Alice got a big smile as she leaned close to him. "You are the most handsome valet I have ever known." She kissed him on the cheek and climbed into the car.

Carl climbed into the car, adjusted his seat, and started the engine. He looked at Alice and winked as he started humming "Jingle Bells," backed the car out of the driveway and drove the Kruger family to church.

—*—

The EG Faith Center was filled to capacity. Their Christmas Eve service was always popular with the community and surrounding area. The church went all out with a live nativity scene, a thirty-six person choir, the Everland Stringed Orchestra, and several key vocalists. Stanley MacKenna Thorne used to be a regular on the program until he developed Alzheimer's years ago and was moved to a facility. Tonight, he made his debut return, and the crowd anxiously awaited.

The room was filled to capacity. More than six hundred people filled the pews and stood along the walls in the back of the large room. Families, children, elderly, singles, men, women, sinners and saints alike crammed into the standing room only sanctuary. The pastor welcomed everyone and started plowing through the program. After the reading of the Christmas Story and a song by the choir, the pastor introduced Stanley. He told briefly his history, his World War Two service, recipient of the Congressional Medal of Honor, and his recent bout with Alzheimer's. The old man walked onto the stage in a tuxedo and wearing a big smile. He barely looked seventy much less in his nineties. The audience of six hundred plus guests offered a standing ovation before Mac even opened his mouth.

After nearly a full minute of applause, the audience quietly seated, the lights dimmed and the orchestra struck the first note. They played 'Cantique de Noël,' known by many as 'O Holy Night,' flawlessly. The music wafted through the large room. The perfect acoustics allowed the sound of the strings to float through the air with grace and magic.

Mac took a deep breath, opened his mouth and began singing the song effortlessly. His voice melded with the orchestra creating an inspiring moment for everyone listening. The performance was sheer perfection.

When Mac finished, the audience in the sanctuary jumped to their feet and burst into applause and cheers. "Bravo! Magnificent! Amen!" People were weeping, laughing, whistling and clapping feverishly. The animals on stage became jittery and had to be held tightly by the nativity cast. Never before had anyone moved an audience at this church like Mac. It was a divine moment for many.

Mac bowed slightly and stood before the crowd … and wept.

—*—

It was Christmas morning at the Howe's house. The presents were piled high under the Christmas tree. The ornaments shimmered, the Christmas lights flickered, and the star that topped the tree blinked. Nine year-old Bonnie was wearing a beanie cap to hide her hair loss from the chemotherapy she was taking. She was seated on the floor with her younger sister and brother in a large circle around the tree. Each child had a stack of toys and clothes in front of them; mostly toys. Bonnie's brother was playing with a large Tonka dump truck and making sounds like an engine, brakes and bells as he moved about picking up smaller toys and dumping them in different locations.

Her younger sister had two dolls propped up on her lap. She impersonated their voices as though they were talking to each other about some event at the theater.

Mom and dad were seated on the sofa soaking in the excitement and events of the morning. They each had a cup of coffee in their hands, sipping, smiling and whispering about their young children. What a joy they were to them, but it pained them to see Bonnie so frail and sick. All they wanted for Christmas was to see their daughter well.

Bonnie surveyed the toys in front of her. They were age appropriate for any girl age ten and up. She stacked the toys in a small pile to the right of her. This was soon to be her discard pile. She glanced over the clothes and thought, *cute*. She examined the fabric, labels, seams and hems, and decided they were nice and worth keeping. They went into a different pile; the "keep" pile. She grabbed one of the toys from under the tree and examined it closely. It was a Rubik's cube.

The Rubik's cube was invented by Erno Rubik in Hungary in 1974. The cube is a six-sided plastic box with nine colored squares on each side. The goal is to move various rows of squares by twisting and turning the cube, moving three squares at a time, until all of the colors on all six sides line up. The cube was mass-produced in 1975 and became one of the most popular toys in the world with more than 300 million cubes sold worldwide. The cube has hundreds of millions of color combinations that frustrated millions of people because they could never solve it. Now, it was Bonnie's turn to try and solve the puzzle.

Bonnie's parents watched their daughter as she examined every side of the cube. She slowly turned it from one side to another. She held it up, turned one row of three squares, stopped and examined the cube.

"I don't think she understands it," her mom whispered. "Maybe we should tell her …"

Bonnie's father interrupted. "No. Let's see if she can figure this out."

They both watched the little girl as she continued to examine the cube. She would turn some squares a couple of times, examine the cube, turn some more squares and examine the cube again. Suddenly, Bonnie turned to her parents and smiled. She never looked down as she started turning the rows of three and repositioning the cube at a lightning pace. Her hands were a blur as she twisted, turned and shifted the little cube in her hands. Her parents were aghast. In less than a minute, she stopped, held up the cube with solid colors on each side and smiled. "Piece of cake," she said.

Her mother let out a 'yelp!' and spilled her coffee.

—*—

"Oh, my gosh. Quinton!" Abbey held up the diamond and gold necklace. "We can't afford this!"

Quinton smiled. "We can, and you are worth every penny, babe. I wanted to surprise you."

Abbey was stunned. "Uh, surprise? Yeah, I'm surprised. I'm … speechless. This is beautiful!"

Quinton reached out and took hold of the stunning necklace. "Here, let me help." He reached around his wife as she held her hair up and clasped the necklace. "There. It looks beautiful. You look beautiful."

Abbey gave him a big hug and kiss and then dashed to the hallway mirror. "Oh, Quinton. This is beautiful." She paused for a second. "How can we afford this?"

Quinton knew if he told her the truth, she would be angry with him. He couldn't tell her that he entered a couple of large poker tournaments at the Indian casino in Jackson on the days he said he was training for the tertiary filtering system at the plant that was being installed. Truth was there was no training on the new filtering system. It was only an excuse that enabled him to apply his extraordinary skills at reading other poker players, calculating the odds, and knowing which hands were the best opportunity to win and how to play them, courtesy of the numerous books he read on the topic without Abbey knowing. "I got a big bonus at work and I wanted to surprise you."

Abbey admired the necklace. She returned to the sofa and hugged him. "You certainly did that, Mr. Quinton Lemolo."

Quinton smiled as he held her and thought about the fortune he would soon be vying for in February at the World Poker Tour in Las Vegas. All he had to do was set up the excuse to be gone for a few days without his young, naïve wife.

Year Two—2015

New Year

January, 2015

CARL KRUGER WAS sitting at a table in the director's office with a small stack of papers in front of him. He picked up the scans and studied them carefully. He switched back to some medical notes and back to the scans. "Remarkable," he mumbled.

"I know," came the reply from across the table. It was Owen Pitke, the Director of the Johns Hopkins Alzheimer's Disease Research Center and the Director of the Division of Cognitive Neuroscience in the Department of Neurology at Johns Hopkins University School of Medicine. Owen was in his mid-sixties with long, white hair along the sides of a large, bald head. He had a terrible comb-over that was obvious. Over the years it grew until it went from one side of his head all the way across the top to touch the hair on the other side. It was trimmed perfectly and greased to lay flat against the scalp. The pitiful thing was everyone could see his scalp through the comb-over, and everyone thought it looked ridiculous, except Owen.

Owen sported a mustache, sparkling blue eyes, and a contagious smile, even with the bad comb-over. His educational and professional experience prepared him for the challenging task of coordinating research teams to attack the dreaded disease with the hope of finding a cure. He had a Ph.D. in neurology with significant education and experience in the area of disease-related cognitive changes with a particular focus on Alzheimer's disease. He drafted numerous papers regarding the relationship of cognitive change to brain structure and function as assessed through imaging. He has viewed thousands of brain function images, including FMRI[7] and PET scans. These, however, were different. "I've never seen anything like this, either. That is why I wanted to bring these straight to you for an individual assessment and discussion of coordinating a research team."

"A research team beyond our current resources?" Carl asked.

"Absolutely." Owen took the scans from Carl and flipped through them until he found the one of Mac's neural activity. "Look at this. This scan tells me the patient's brain is near normal. You and I know moderate to advanced Alzheimer's atrophies the brain—physically—while it destroys synaptic activity. These markers indicate the brain is regenerating from some type of neurogenesis."

"Impossible," Carl said as he stared at the scan.

Owen continued, flipping to another scan. "And this. This is from a middle-aged man who suffered from significant migraines."

"Familial Hemiplegic Migraine?"

"No. The diagnosis is Sporadic Hemiplegic Migraine. He has photopsia[8] and minor motor weakness. What is remarkable in this case, the patient developed a photographic memory in less than six months. His scan shows neural activity significantly above average in the medial temporal lobe and hippocampus regions."

"The other areas also show significant activity."

Owen pulled out the last scan. "And then we have Bonnie Howe." He laid the scan in front of Carl. "A nine year-old girl with leukemia who, before last summer, was average intelligence." Owen laid another page in front of Carl. "Her IQ is now one hundred and fifty-six."

---

7 Functional Magnetic Resonance Imaging—fMRI) relies on the paramagnetic properties of oxygenated and deoxygenated hemoglobin to see images of changing blood flow in the brain associated with neural activity.

8 Photopsia—A disturbance of vision consisting often of unformed flashes of white and/or black or rarely of multicolored lights.

"One fifty-six? Einstein was one-sixty. How?"

"That's what I want you to find out; why we have three patients of three different age groups who, last spring, were common, everyday people. Now, their brain activity is off the charts. Why? What changed? What is different? What made them the way they are?"

Carl leaned back in his chair. "How can anyone with Alzheimer's regain memory?"

"If I had that answer, we'd all be rich," Owen said.

"Unbelievable." Carl said. "When do we start?"

Owen started gathering the scans and other documents as he spoke. "We have arranged for the patients to arrive the first week of February for three days. You will have unbridled access to run any and all tests you deem necessary short of invasive procedures. I'll go over the specifics tomorrow. I'll need to have you decide if you want to add anyone to your team for this project."

"Anyone?"

"Anyone. Carl, we will have one shot at this. We are fortunate one of the neurologists was from Johns Hopkins and contacted us to do further research on their patients. This is our only shot, so we need to make it a home run."

Carl smiled. "It could be a grand slam on this one."

—*—

It was more than a week ago when Carl and Owen had their initial meeting to discuss the three patients from California. Since then, Carl had multiple meetings with his key staff to discuss the tests to run; toxicology, genetics, pathogens, environmental, scans, motor functions, memory; everything. They discussed what samples they could obtain with non-invasive procedures. Nothing was left to chance. This was going to be the most extensive testing of any subject that Johns Hopkins University ever conducted.

Carl was pleased that he had a team of professionals able to analyze the samples with the best equipment possible. He had doctors and scientists in every area of expertise. But there was one person he wanted on the team more than anyone, regardless if his current members were capable.

"Imar!!" Carl walked briskly to the edge of the walkway next to the TSA[9] officer guarding the exit. "Imar!!" he yelled as he waved his arms. "Over here!"

An elderly man with wild, gray hair turned towards Carl and smiled. "Carl," he said softly, aware that Carl could not hear him. Imar waved his hand. "Good to see you," he whispered.

The man, who looked remarkably like Albert Einstein, walked to the exit to greet Carl. As Carl shook his hand, Imar placed his other hand on top and looked Carl in the eyes. "Carl. My good friend."

"Imar. It is so good to see you." Carl pulled Imar into a hug. The older man was shorter than Carl and looked uncomfortable as his head was pressed into Carl's armpit. "New deodorant?" Imar asked with his face squished.

Carl stopped the hug and laughed. "Yes, new deodorant." Carl took the small suitcase from Imar. "Let's get your other bags and we'll head to the hotel."

"If you don't mind," Imar said with his slight Austrian accent, "I'd rather go to the lab and see what you have going."

Imar Spaan was a world renowned microbiologist. He won the Nobel Prize in 1996 jointly with Peter Doherty and Rolf Zinkernagel for their discoveries concerning the specificity of the cell mediated immune defense. He was a brilliant scientist, and Carl had every opportunity to observe Imar under pressure when they worked on the 19Q project together. Imar was instrumental in identifying the cause of the chromosome mutation and subsequent deaths of the recipients. Carl respected his opinion, was impressed with his abilities and trusted the man implicitly. They developed a lasting friendship while working together on the project.

The two men continued to chat as they approached the baggage claim turnstile. "The director of the institute has agreed to pay for an apartment for you during your entire stay. Wouldn't you rather go there first? I know it's been a long flight and you've been up most of the night." Carl asked.

"An apartment? How long do you expect this to take?" Imar asked.

"No idea, Imar. This is new ground."

---

9 Transportation Security Agency

"But I only brought clothes to last a week."

"Then we will have to go shopping tomorrow." Carl grabbed Imar's bag off the baggage claim turnstile. "So, to the apartment?"

"No. I'd much rather go to the lab."

"That's fine. We can talk on the way."

—*—

Alice Kruger entered the lab on the far side of the room. Some of the analysts and doctors knew her from her frequent visits to the facility to pick up Carl. Her interests in research never waned, even though she chose to leave her career at the Centers for Disease Control, marry Carl and move to Maryland. Alice convinced herself that her role as a mother was just as challenging as a scientist. At times it certainly was. But she missed the camaraderie of the people and the adventure of a research lab. It was for the best, however.

Alice's physical limitations were something she could manage. She waddled her way through the lab on her crutches with her signature messenger bag around her shoulders and a smile on her face as she tried to sneak up to Imar from behind.

Imar was discussing the classification of bacteria and viruses with two analysts when he heard Alice's crutches clinking as she approached. He turned around to see her. "Alice!!" Imar walked briskly across the room to greet her.

"Imar, you crazy old scientist," she said.

Imar reached out and hugged Alice, nearly lifting her and her crutches off the floor.

"I was wondering when I would get to see you, Imar," she said. "It has been so long."

Imar smiled. "It has. You look beautiful!"

"Thank you, Doctor."

"How did you know I was here? In the lab?"

"Carl sent me a text. I knew you were coming in and I knew you well enough to know you would want to come here first. Carl just confirmed it."

Imar looked at Carl who just entered the room. "Well, I'm glad he did," he said as he turned back to Alice and smiled. "You look good, Alice," Imar said with a sincere smile.

"As do you." Alice leaned forward and gave Imar a kiss on the cheek just as Carl walked up. "Alice. Who is that man you're kissing?" he laughed.

Alice smiled back. "Jealous?" She took hold of Imar's arm. "This is my date."

"Date?" Imar said. "Date? I wasn't aware of that proposal."

"You are now," Alice said. "I've always been attracted to brilliant men. Can you two gentlemen go to lunch, or are you so busy you can't leave?"

"Lunch, definitely," Imar said.

"Lunch it is," Carl agreed, and the three of them headed out the door with Alice holding onto Imar's arm and Carl carrying one of her crutches.

Discovery

February 4–6

THE LARGE CONFERENCE room at the Alzheimer's Disease Research Center was packed. A long, wooden table was positioned in the middle of the room with Stanley MacKenna Thorne, Bonnie Howe and her parents, Quinton Lemolo, Owen Pitke, Imar Spaan, and several analysts. Bonnie was wearing her signature knit beanie and sitting at the end of the table. Carl Kruger was standing at the other end with a large whiteboard that doubled as a projection screen.

"Thank you all for being here today." Carl nodded to the three patients. "The three of you have recently had tests performed that revealed something quite unusual in the field of medicine: elevated neurological activity. Each of you has experienced some form of change in your memory retention and cognitive abilities over the past year that is unexplained. The goal of this project is to conduct a series of tests, obtain samples and run analysis on each of you to see if there is a common thread that links the three of you together."

"Other than we all come from the same area?" Quinton asked.

"Yes. Other than you all came from the same area, which is one of the factors," Carl replied. "We will be conducting scans, x-rays, extracting blood and fluid samples, conducting written, verbal, and spatial tests, and more. It's going to be extensive, so please bear with us. My team will work as expeditiously as possible."

"We didn't want to take Bonnie out of school for too long. She needs to stay connected to her classmates and studies," Bonnie's mother said.

Bonnie rolled her eyes and stared at the papers in front of her.

"It will take us three days or less to run all of the tests and gather samples. We expect you will be on your return flights this weekend," Owen said.

Quinton smiled. "Good, because I need to make it to the World Poker Tournament in two weeks. That is one event I do not want to miss."

"I'm sure this project will not cause you any delay, Mr. Lemolo," Imar said.

"I trust the rooms are to your liking?" Carl asked.

"Pretty nice," Quinton replied.

"Great. We are also providing your meals and transportation during your brief stay," Carl said. "We have prepared a welcome packet that each of you have received. It contains a map of the facility and area, meal tickets, transportation contacts for your down times, and a research schedule."

"Down time?" Bonnie's mother asked.

"Yes. There will be times when you will have several hours off. You are welcomed to visit the area, take in a movie or museum, maybe go shopping. We only ask that you be aware of the time so you can be at your scheduled tests on time. Your driver will also watch the time and keep you on track. We will also give you a pager to carry during test hours."

"Driver, eh?" Mac said. "I think I'm going to like this," he said as he rubbed his hands together.

"We hope you do," Owen said.

"Our first series of tests begin in one hour. We have staff to take you to your rooms so you can review the information and prepare for the tests. Please report to your first test station by 10:00 A.M. Are there any questions?"

Mac cautiously raised his hand. "So, you just want to test us and let us go on our way. That's it?"

"Well," Owen straightened his hair. "… we will be in touch with you weekly through email or phone calls to see how you are progressing. We don't want this to be intrusive or burdensome, but something extraordinary is happening to each of you and we want to see how it develops over time."

Carl surveyed the room looking for any cue of puzzlement or confusion on the faces of the guests. In stark contrast, they seemed bored. "O.K., then, if there are no more questions, we will see you at ten."

—*—

Bonnie Howe was seated at the table with Kim, an analyst on the Project team. The room was small with blank walls except for a picture of a waterfall. There were no windows and only one door. Bonnie fidgeted with her beanie, pulling it almost over her eyes.

"Bonnie, my name is Kim. I'm going to conduct a memory test with you."

"With me or on me?" Bonnie asked.

"Well, on you would be more appropriate to say." Kim had a folder in front of her that she opened. "I have a story here that I would like you to read to yourself." Kim handed the pages to Bonnie. "Take your time and read every word."

Kim watched as Bonnie scanned the paper. After less than two minutes, Bonnie looked up at Kim. "Done," and handed the papers back.

"O.K. Now, can you tell me what was written on the paper?"

Bonnie let out a deep sigh and rolled her eyes. She acted bored.

"If you don't remember anything, that's O.K. We can ..."

"Abraham Lincoln was born February 12, 1809, the second child of Thomas Lincoln and Nancy Lincoln, in a one-room log cabin on the Sinking Spring Farm in Hardin County, Kentucky. Lincoln's paternal grandfather and namesake, Abraham, had moved his family from Virginia to Jefferson County, Kentucky, where he was ambushed and killed in an Indian raid in 1786, with his children, including Lincoln's father Thomas, looking on. Thomas was left to make his own way on ...."

Kim listened and followed along as Bonnie recited the story of Abraham Lincoln verbatim. She waited to mark errors in red, but was stunned to hear the little girl recite the entire, two-page story flawlessly. One thousand forty-eight words, no errors.

—*—

Mac laid on the table in the examination room. He stared at the lights above him. He was dressed in a gown wearing only his boxers and socks. "Do you have a blanket, Doc? It's kind of chilly."

"Sure." The young lady opened a blanket and spread it over him. "There ya go. And, just to be clear, I'm not a doctor; just an analyst."

"I was 'just a soldier' when I was in the war. You should take greater pride in what you do," he said.

The aide smiled and continued to hook up wires and receptors on Mac's chest, arms and legs. She examined the connections, punched a few buttons on the machine and wrote some notes on a pad. Then she entered some information on the computer monitor. "There. I think we are ready."

"Let her rip."

"O.K. I need to have you step over here to the treadmill." The aide held Mac's elbow as she helped him over to the treadmill. "Now, I'm going to start off slowly. Then, as we progress, the treadmill will go incrementally faster. If you feel it is going too fast, tell me to stop. I can slow it down for you. Understand?"

"Sure. Let's go for a walk," Mac said as he smiled at the young lady.

The aide turned on the treadmill and watched the monitor as Mac started walking. Several lines appeared on the screen as she toggled from screen to screen. Mac started humming 'I'm Walking,' a song recorded by Fats Domino in 1957.

The treadmill whirred as Mac walked and hummed.

—*—

Carl Kruger was standing behind the analyst watching him enter the program data into the computer. Quinton Lemolo was lying on a table in front of the large cylinder. He was observing the lights and a picture of a meadow taped to the ceiling. "Can you hear me Quinton?" the analyst asked.

"Yep."

"Good."

Carl walked around the partition and stood next to the table. "I know you have had this procedure done before, but I wanted to let you know what to expect. The machine will soon start up and make a few noises. We will place a pair of earplugs in your ears then slide your table into the cylinder. We will be taking pictures of various layers of your brain as you lay still and breathe slowly. This first round will last about ten minutes. Then, we will take a break for a few minutes as we review the scans. After the break, we will conduct scans on another part of your body. We will continue the series of scans and breaks until we have scans of your entire body."

"This could take a few hours, right?" he asked.

"It will take some time. If, at any time, you feel like you need to stop, we can have other tests run while you recover from this. The MRI is a bit daunting because of the loud noise it creates. It almost sounds like a rapid jackhammer going off right next to your head."

"I know, Doc," Quinton said, showing he was annoyed. "I've been through this once already."

"I understand. I just wanted to make sure you did, too. Any questions?"

"Nope. Let's roll," Quinton said as he snapped his fingers.

Carl walked back behind the glass partition and said something to the analyst. The machine started whirring as the table slowly slid Quinton into the tube head-first. The analyst directed Quinton to lay still and breathe slowly. The machine started a rapid hammering sound that was deafening. The MRI scan had begun.

—*—

Bonnie Howe sat in a chair, quite comfortable and very interested in the technology she was observing. A large, rectangular box sat on a table connected to several monitors. Two large black cables ran from the box to a belt that had thirty-two colored wires coming out of it. A round receptor was attached to the end of each wire.

The doctor walked in as Bonnie was observing the equipment. "Hi. My name is Molly." The doctor gently shook Bonnie's small hand. "We are going to run a test that will take pictures of your brain as we show you different things, okay?"

Bonnie looked at the doctor. She was young and attractive, much like Bonnie's new teacher. "Sure. I guess I don't really have a choice, do I?" she asked.

The doctor was taken aback by the comment. "Uh, we always have a choice, Bonnie. I thought you and your parents wanted these tests run?"

"They do more than me. But I'm O.K. with it. It's kind of interesting." Bonnie looked at the equipment as the doctor prepared the headset.

"I'm glad that a young girl like you is interested in technology like this. Do you know what you want to be when you grow up?" the doctor asked as she continued to work with the headset.

Bonnie looked at the doctor. "I just want to grow up," she said and then turned back to the equipment.

The doctor never stopped working with the wires and headset as she continued the conversation. "So, do you think you won't because of the leukemia?"

Bonnie pulled her beanie tight and looked down at her feet. "Yes."

"Well, I believe you will. The research and the treatment of leukemia have provided terrific breakthroughs, especially for a young girl like you," the doctor said as she smiled at Bonnie.

"Great," Bonnie said quietly. "So, what type of imaging system is this?" she asked, changing the uncomfortable topic.

"It's a Functional Electrical Impedance Tomography Unit," she answered.

"So, it's going to take readings of the electrical impulses of the brain through this headset and create pictures from it, correct?" Bonnie asked.

The doctor stopped and looked at Bonnie. "Close. It's going to put small electrical currents into your brain through some of the electrodes while the other electrodes read the electrical signals from your brain and create an image of activity." The doctor paused. "Have you had this test before?"

"No." Bonnie paused. "Is it going to hurt? The electricity?" she asked.

"No. Not at all," the doctor reassured her. She continued to turn some knobs and flip a few switches. She watched the computer screens change each time she made an adjustment. "There. We are almost ready." She turned to Bonnie. "I need to place all of these receptors on your head in various areas. To do that, I need to clip the hair short in specific areas so the receptors will stay in place."

"I know. That's O.K.," Bonnie said as she pulled her beanie off, revealing her short hair. "I lost most of my hair last year with my treatment, but it's starting to grow back. I expect I will lose it again. So, it's no big thing."

"Is that why your hair is short now?" the doctor asked.

"Yeah. I was just getting to like it."

"Well, I'll make really, really small clips and keep the hair kind of long where I can so you won't even know the hair is missing. We don't need to take it to bare skin; just shorten the hair a little more." The doctor could see that Bonnie was concerned and downcast. She smiled as she took Bonnie's hand. Bonnie looked at her green eyes as the doctor continued. "I believe your hair will grow back, that the doctors will stop your leukemia and you will grow into a fine young lady."

Bonnie smiled at the doctor. "Thank you, Molly." She was sincere and sweet, then pulled her hand away and quickly changed to her analytical self. "How long will this test take?"

"About an hour." The doctor took a pair of scissors and slowly, methodically clipped small sections of hair. She explained the procedure as she spoke. "I'm going to place these receptors all around the base of your brain and across the top of your head … here … and here. Then I'm going to run one series of tests to create a starting point for a picture of your brain. After that, I'll give you things to look at, sounds to hear, stuff like that, and keep taking pictures to see how your brain reacts." The doctor continued to clip away at small sections of Bonnie's hair. "You won't feel a thing, except the wires pulling on the receptors."

"Sounds really interesting. Will I get to see the pictures when we are done?" Bonnie sat still just like she was getting a haircut.

"I think we can work something out. I'll see what I can do." The doctor laid the clippers down and looked at Bonnie's hair. "There, all done. Do you want to see?"

"Sure."

The doctor reached in a drawer and grabbed a hand mirror and gave it to Bonnie. Bonnie lifted sections of her hair as she gazed in the mirror. "Nice job, Molly."

"Why, thank you." The doctor smiled and took hold of the headset. "Now, the bonnet." She slowly attached one colored receptor at a time while Bonnie watched in the mirror. The doctor would pull the paper off the receptor, lift a section of hair and press the receptor onto the shortened hair. She continued this process until the receptors covered Bonnie's head. "There. All done. What do you think?"

Bonnie studied the wires and receptors in the mirror. "I think I'd rather wear a beanie."

Molly laughed and Bonnie giggled.

Bonnie had a friend.

—*—

"You want me to do what?" Mac asked, standing at the counter looking at the plastic containers lined up along the countertop.

The middle-aged analyst looked annoyed. She seldom had people question such a simple request. "I want you to go in the restroom, there," as she pointed to the door and continued. "... and obtain a urine sample in this." She held the plastic cup up to Mac's eyes so he could easily see it. "Then, I want you to take this packet to your room with you and obtain a stool sample by tomorrow morning. It has instructions inside." She held the packet and plastic cup in front of Mac's face.

"That's what I thought you said," Mac said as he snapped the containers from the woman. He paused for a few seconds, examined the containers and continued. "And just what will these samples tell you about my brain?" he asked.

"Probably nothing," she replied.

"I hope to shout," Mac said as he walked into the restroom, obviously irritated.

—*—

Quinton lay on the table staring at the instruments on the metal tray. His heart started pounding as he examined the long needle and syringe. His migraines were less frequent, thanks to the medication he was taking, but he wanted them to stop completely. They were extremely distracting, especially when he was in a tournament and had to focus the entire time. He couldn't just walk away and allow his competitors to gain an edge. He had to stay in the game, stay on task. That was why he chose to be the only volunteer for the spinal tap.

"Are you feeling all right?" Carl asked.

"Yeah. A little nervous I guess. Especially after seeing that needle," Quinton replied as he pointed to the long needle and syringe lying on the tray. "Haven't seen one that big since I worked on my daddy's ranch. And it wasn't for a man, either."

"I understand," Carl assured him. "Doctor Savoie is an excellent neurosurgeon," Carl said as he nodded to the middle-aged woman preparing the instruments for the procedure. Carl continued. "She has done many of these and will be gentle, I assure you."

"I will," the woman confirmed. "I'd like to explain the procedure to you again, even though someone informed you when you signed up."

"Sure," Quinton replied.

"First, you will lie on your left side and curl into a fetal position with your neck bent down as much as possible. Try and touch your chin to your chest. I will prepare the area around the lower back using an antiseptic. Then I will push on the area to locate the entry point in the lumbar region. Once the appropriate location is palpated, I'll insert a local anesthetic under the skin and then along the intended path of the spinal needle. That will sting until the anesthetic takes effect. Once the area is numbed, I will insert a spinal needle between the lumbar vertebrae L3/L4 or L4/L5 and push it in until there is a 'give' that indicates the needle is past the ligamentum flavum. You will feel pressure as I push. Don't move. Just try and relax and stay fixed. Once I pass the first layer, I will push the needle again until there is a second give. Then, I will withdraw the stylet from the needle and extract the spinal fluid. Any questions so far?"

"No. Sounds like torture," Quinton said.

"Sounds like it but, hopefully, isn't," Doctor Savoie replied. Her tone was subdued and comforting. "The size of the needle is what scares most people. The procedure has been modified over the years to inflict as little pain and discomfort as possible." She paused for a second. "There are some possible side effects that could occur from the procedure. The most common is a spinal headache. Since you have suffered from migraines, the likelihood of receiving a spinal headache is high. That's why I have chosen to give you an intravenous injection of caffeine."

Quinton looked surprised. "Caffeine? You mean like a coffee?"

The doctor smiled. "I do. Studies have shown that a caffeine IV is effective in aborting these spinal headaches."

"Terrific," Quinton said. "Make mine a marble mocha macchiato."

Everyone chuckled at the comment.

"I'll try," Doctor Savoie said. She continued with the explanation of the procedure. "After we inject the marble mocha macchiato, I'll have you lay on your back for approximately two hours. This allows the cerebral fluid to rebalance in your system. You will need to lie still, so I suggest you take a nap or something. We have music that you can listen to, or watch the TV on the ceiling."

"O.K. I'm ready for a nap anyway."

Doctor Savoie placed her hand on Quinton's shoulder. "O.K., then. Ready to begin?"

Quinton took a deep breath. "Sure. I need that coffee."

—*—

The sports bar was noisy. People were eating dinner at several tables while multiple televisions displayed several sporting events. Imar Spaan, Carl Kruger and Owen Pitke were sitting at a table enjoying their drinks and relaxing. The tests on the patients were complete. Over the last three days their team took scans, tested mobility, drew blood, took samples of urine and stools, checked the eyes, ears, skeleton and brain; every conceivable, non-invasive test possible that they could conduct in the short timeframe was applied. The three doctors reviewed the list of tests completed to make sure nothing was missed. They were very pleased with themselves and were enjoying the relaxed wrap-up conference at the sports bar.

"Nice job, gentlemen," Owen said as he lifted his draft beer for a toast. "Carl, you did a great job of coordinating all of these tests. I know the logistics of some of these was not easy considering the location of the facilities with the equipment. You need to be commended."

The men clinked glasses. Imar leaned forward to address the director with his Austrian accent, bushy mustache and wild hair. He had a little draft beer on his mustache that made him look even crazier. "So, does that mean you are giving Carl a raise?" he said as he raised his bushy eyebrows.

"Imar!" Carl yelled.

Imar laughed. "How do you say, 'Gotcha?'?" He laughed again, this time with Owen and Carl joining in. "Doctor Pitke. As you can see, I am a big fan of Mr. Kruger. He is worthy of a raise, regardless of what you are paying him now. What I am saying is, the work he does is … priceless."

Owen smiled. "Doctor Spaan. I would have to agree." He looked at Carl and winked. "It is good to know that Doctor Kruger has agreed to continue to volunteer here for the next year."

Imar was shocked. "Volunteer?" He turned to Carl. "Volunteer?"

Owen and Carl both laughed as Carl slapped Imar on the shoulder and said, "Gotcha back."

The men laughed heartily as a waitress brought a plate of wings for them to enjoy. Their camaraderie was refreshing to Carl. He missed Imar, and respected the brilliant scientist and doctor greatly. "It is going to be fun working with you on this project, Imar," Carl said.

Imar wiped his forehead and mustache with a napkin. "Yes, fun and difficult, no doubt. I suspect the work is only starting, would you agree Doctor Pitke?" Imar took a hearty bite of chicken wing, leaving a blob of barbecue sauce on his mustache.

"I would. We have a good start on gathering the data and samples. Now, we need to do the detailed analysis of everything to see if there is causation for their strange ability."

Carl responded first. "I believe there is."

"As do I," Imar agreed. "The coincidence of three people having the same type of rapid neural development from a geographic area at the same time is highly suspect."

"It is," Owen replied. "The next step is to analyze all of the samples and tests. I have a list in my office of the various areas of concern and what research to conduct. I would like the two of you to review the list over the weekend with your key doctors, Carl, and assign sections as appropriate. I'm going to call a meeting Monday with everyone to establish a project timeline with responsibilities and checkpoints. I don't want anything missed."

"Agreed," Carl said.

"Very well, then. It appears we are ready to go with the project," Owen said as he rose and started to leave.

Imar reached out and stopped him. "I believe we should name the project with something appropriate to their condition," Imar suggested.

"We could call it 'The Elephant Project' because of their memories," Carl suggested. Everyone laughed. "Well, it is going to be a *big* project," Carl continued and they laughed again.

"I don't think these geniuses would appreciate being compared to an elephant," Owen said as he took his seat again.

The room became quiet as the three men thought. Suddenly, Imar sat straight up in his chair. "That's it," he said as he wiped the barbecue sauce from his fingers and mustache, only removing some of it. "It's simple. Memory. 'The Memory Project'," he said as he raised his hand. He smiled widely showing pieces of chicken caught in his teeth. He looked like a wild man.

Owen looked at Carl and back to Imar. "But we've confirmed increased IQ in the girl."

"Doesn't matter, Carl. We are seeing memory improvement. That's the key." Owen said. He raised his near empty mug for a toast. "The 'Memory Project' it is." The men clinked mugs and toasted the name of the project. Owen wiped his mouth, checked his hair and rose. "Well, gentlemen. I suppose I shall see you Monday for the first official meeting of The Memory Project."

Owen left Carl and Imar in the sports bar to reminisce about the 19Q project. They both agreed that this project was much different, yet oddly similar to 19Q. They hoped it would be much more successful.

February 19

THE LAB AT the JHU Alzheimer's Disease Research Center was 'state of the art.' Several analysts and scientists were seated at tables and counters working on a myriad of samples that had been collected less than two weeks prior. They were viewing microorganisms through microscopes, pictures of viruses on a monitor, running centrifuges, DNA analyzers, viewing x-rays and scans and charts and graphs. Everyone had a checklist and an assignment and was working efficiently and quickly. There was no time for horseplay. The stakes were high and everyone knew it. Yet nothing definitive had surfaced.

Imar Spaan was coordinating the microorganism and cellular research of the samples. The task was daunting. He estimated the research could take months to complete. Each person has thousands of microorganisms in and on them at all times. To determine something unusual or abnormally present in a person as compared to the general population is akin to looking for a needle in a haystack.

His first task was to analyze the blood from the patients to determine if there was an increase in antibodies as a result of prior or current infections. All of the samples revealed a level of the IgG antibody indicating each person had an infection in the recent past. The results were problematic for the sample drawn from Bonnie Howe since she had leukemia that was in remission, which they already knew about. Quinton had a case of measles and some skeletal issues, nothing significant. Mac had hepatitis from the military, and once had malaria, which they knew about. Nothing new jumped out to the team. Everything looked to be within normal limits. Yet, it was conclusive that there was something that caused an infection of some type in the past for each patient. Something.

The next step was to screen for both bacterial and viral agents that could have been the cause of the prior infection. The blood tests were negative. Nothing out of the norm surfaced.

Imar decided to test the cerebrospinal fluid extracted from Quinton on his own. Since his task was to find a common link between the three patients and not just one, he delayed the spinal fluid research until he was free enough to conduct the tests at his own pace. Testing Quinton's spinal fluid was in contrast to all of the other tests. If anything was present, it would be from a single source and not the entire test group. Validating the findings would be impossible unless all of the patients agreed to a similar invasive procedure to extract the fluid. Even so, Imar was a scientist. Any analytical information from any sample could be a compelling reason to investigate further. It was chasing rabbit trails. He was adept at such research and exhibited it well when he worked with the 19Q team and the Nobel Prize team years ago. That was why he decided to personally analyze the fluid while his team continued to work on the other samples.

The first test he conducted was a PCR[10] screen of the fluid. Imar carefully opened the container holding Quinton's spinal fluid and extracted a few drops with a long pipette and carefully dropped them into small vials. Then, he added a drop of green dye to one sample, red dye to another, and clear fluids to the others. He placed the strip of vials into a PCR machine, pushed a few buttons, and walked away.

—*—

Sacramento International Airport was busy. Cars were double parked along the departure gates as people pulled travel bags and suitcases out of the trunks, kissed, hugged and walked into the terminal usually with their drivers waving 'bye.' Quinton Lemolo was just another passenger on another trip, just like everyone else.

---

10 Polymerase Chain Reaction (PCR) is a scientific technique in molecular biology to amplify a single or a few copies of a piece of DNA across several orders of magnitude often used for the detection and diagnosis of infectious diseases.

"Okay, hon. Have a great trip," Abbey said as she leaned into her husband and kissed him. "Be sure to call me every night so I can think of you before I fall asleep."

"I will, babe. You can count on that," Quinton said as he kissed his wife again. "I'll be back Sunday night." He leaned into the car to kiss his baby girl. "You be good for your mama, sweetie," he said as he kissed her again.

"I'll be here to get you. I've got your itinerary." Abbey closed the trunk and continued. "Please don't gamble, Quint."

"I won't. I promise." Quinton had become an excellent liar. Having the ability to remember everything enabled him to remember even the smallest details of a lie. Most liars are discovered when they can't remember a piece of a lie and get caught in the lie by restating something incorrectly. Not Quinton. He could remember every detail of everything, so lying was simple; easy.

"Good. Are you going to take in a show or anything? I mean, you're going to be in Vegas. They have great shows, you know."

Quinton opened the handle on his bag. "I don't know. We're going to be pretty busy with this class. The only reason they chose Vegas was the low cost." Quinton walked back over to his wife. "I wish you could come, too, but, to be honest, I'm not going to have much free time. I'm gonna be busier than a raccoon in tall corn."

"Well, I hope you have a little free time. I'll see you Sunday, Quint."

The couple kissed again and Abbey climbed into the car and drove away as Quinton walked into the terminal with scores of travelers. He was on his way to the World Poker Tournament in Las Vegas to vie for the four million dollar prize.

—*—

The PCR machine was printing a report as Imar stood next to it. When it finished printing, he picked up the papers and slowly scanned through them. Carl Kruger was standing next to him looking over his shoulder. Carl had extensive experience in genetics and was anxious to see what the PCR analysis would reveal.

Suddenly, Imar stopped scanning the documents. "Look at this," he said as he held the paper. Carl moved in closer, then took the document and examined it carefully. "Something is there. This clearly shows a separation of DNA."

"That is likely a virus of some type," Imar said.

"I agree," Carl replied. "Let's run a full spectrum analysis on this fluid right away."

"I'll get right to it," Imar said as he turned and left, leaving Carl standing by the PCR machine alone, examining the report.

February 20

THE SMALL AUDITORIUM was packed. Owen Pitke, Alice Kruger and the analysts and doctors working on the project were present in the audience. Carl Kruger and Imar Spaan were on the small stage backed by a large screen with a terminal next to them and a stack of documents behind them. The audience was murmuring, excited with the anticipation of the announcement that something had been discovered. They were eager to see what it was and excited to be a part of a discovery that would likely be monumental in the cure for Alzheimer's disease.

"Thank you for being here, everyone," Carl began. "Doctor Spaan and I wanted to inform you of a discovery that we," Carl opened his arms to include the audience and continued, "… have made. We have identified a virus that is rampant in the cerebrospinal fluid of one of the patients."

One of the doctors in the audience spoke up. "One of the patients?"

"Yes," Carl replied. "We extracted cerebrospinal fluid from Quinton Lemolo and no other patient. He was the only one to permit the invasive procedure. The virus is rampant in his spinal fluid." Carl moved through some slides and displayed the virus.

Imar pointed to the screen. "What you see here is a type of polyomavirus," he said. "The word 'polyoma' can be broken down into two words; poly for multiple and oma for tumors. It is a somewhat common virus. There are nine known types of polyomaviruses that can infect humans and are believed to be the cause for Merkel Cell Carcinoma, transplant-associated dysplasia, or TSV, and other diseases." Imar advanced the slide to the list of diseases caused by the virus, and continued. "One strand of the virus is the JC virus, named after John Cunningham who discovered it in 1971. This strain is responsible for progressive multifocal leukoencephalopathy or PML[11], a significant disease of the brain. It's a double stranded DNA based virus that typically infects a host with a compromised immune system."

"Such as leukemia or hepatitis, as is the case with the other two patients," Carl said. "However, this sample comes from Mr. Lemolo, who does not have a compromised immune system at present."

Imar continued. "That is true, Doctor Kruger. He does not. This virus exists in seventy to ninety percent of the population already. It is thought the typical route of infection from this virus is through contaminated water of some type since it is found in high concentrations of urban sewage worldwide." Imar paused. "Mr. Lemolo works at a sewage treatment plant in Elk Grove, California."

One of the doctors stood. "But, Doctor Spaan, if this is a typical virus that most people have, why is his employment at the sewage treatment plant significant?"

Imar advanced the slide again. "Because this strain has a DNA marker that varies slightly from the JC virus. It's a new strain of the JC virus. We believe it originated either at the plant or in that area."

A few people in the audience started to mumble.

Carl continued. "To summarize this report, we have a new strain of a virus that resembles a known virus that causes destruction of the white matter in the brain. It is thought to be present in urban sewage systems where this patient works. What we don't know is how he contracted the virus, where he contracted it, if the other patients have it, and what the virus does." Carl paused. "We still have a lot of work to do, but this appears to be our best lead so far. This needs to be our point of focus."

Owen stood and addressed the audience. "I agree. Thank you doctors." The audience applauded. Owen continued. "I'd like to meet with the team leaders in my office in fifteen minutes to discuss the next steps. The rest of you can take a little vacation until next Monday. Thank you all."

---

11 Progressive Multifocal Leukoencephalopathy—a rare and usually fatal viral disease that is characterized by progressive damage or inflammation of the white matter of the brain at multiple locations.

Carl and Imar watched the audience disband. Alice and Owen walked over to the pair and shook hands. "Brings back old memories," Alice said.

"Or nightmares," Carl retorted. "Depends on how you look at it, I guess."

"We need to get cerebrospinal samples from the other two patients," Owen said. "Somehow, we need to get them to agree to the procedure so we can either confirm the findings of the virus as a catalyst, or dismiss it."

"I agree," Imar said. "Until then we should continue to examine the virus and the samples relative to the virus."

"I'll conduct the genetic analysis on the virus at the ICE[12] lab," Carl said. "It will feel good to go back to my old stomping grounds."

"I bet it will," Alice said.

"Okay, then. Carl, work on the genetics of the virus. Imar, the other tests. Alice, lunch." Everyone laughed.

"What about you?" Imar asked.

Owen's smile vanished as he became serious. "After our team leader meeting, I'm going to convince a nine year-old girl and a ninety year-old man to let someone insert a four inch needle into their backs."

---

12 Institute for Cell Engineering

February 21

THE ROOM WAS filled with people. The gallery was laid out like a small basketball arena with people seated in rows, each row elevated slightly above the row in front of them to provide an optimal view of the two premier tables. Tomorrow, these will be the final two tables of the championship and the focus of everyone's attention. Today, however, the field continued to winnow away.

The last two days were grueling. Nearly two thousand entrants for the poker tournament played almost non-stop. It was a small buy-in for only twenty-five hundred dollars. It was one of many tournaments played over the weekend. The event was being recorded for broadcast next season, and Quinton knew that. He had to disguise his looks slightly to avoid detection in case someone recognized him at the event or the broadcast. He had to make sure he didn't draw too much attention to himself, yet he also wanted to place high in the money. Not an even balance to achieve when luck can change everything in Texas Hold'em poker.

Quinton was seated at one of the two tables wearing sunglasses and a baseball cap that read, "WPT-2015." He had a large stack of chips in front of him. He was the man to beat at that table and one of the chip leaders in the tournament. He was concerned that he may have gone too far and had too many chips. People were talking about him and he could see the camera on him at times. It was making him nervous.

Quinton observed everything. He played with some of the players at other tables and knew their 'tell' signs; mannerisms that they repeated depending on the strength of their hole cards and the situation. Quinton watched people at other tables, observed when they won, listened to what was said. He was gathering information by the trainload and no one knew it. He could remember everything.

Quinton watched the dealer shuffle the cards and prepare the deal. He was in the big blind and had to place two thousand dollars as the bet. The cards were dealt. Quinton watched each player look at their cards. The first player, a young man wearing a hoodie and sunglasses with earbuds in his ears, glanced at the cards and laid them down. Quinton watched as he licked his lips. *O.K., he has a middle pair,* he thought. Another tell, easily observed. Quinton could remember the past eight times over the last two days when the young man did exactly the same thing and had a middle pair. Quinton didn't bother watching the other players. He was satisfied with setting himself up to lose on this hand and he wanted to see the young man win. Quinton planned to lose big over the next six or seven hands and eventually lose the tournament. It was near the end of the day and he was already in the money; well into the money. If he lost now, he would earn twenty-four thousand dollars for twenty-eighth place. Not bad for a week-end's work.

Quinton didn't watch the dealer as she laid the "flop[13]" onto the table. Quinton was watching the young man whose lips parted slightly. *He hit!* Quinton knew the man had three of a kind now.

When the betting got to the young man, he checked[14]. Quinton watched closely as the other players bet. He joined in by calling[15] the bets and continued to watch the young player.

When the final cards were on the table, the young player went all-in[16]. Quinton watched as each player folded their cards and dropped out. When the bet got to him, he didn't care what he had. It didn't matter what the young man had. Quinton just wanted to verify his assumptions that the young man had three of a kind. He enjoyed being right. He knew he was good, very good. He called the bet. The young man turned over his cards. The dealer said, "Three eights."

---

13 Flop—In Texas Hold'em poker, the dealer places three cards face up that all players can use in their hand.
14 Checked—When a player does not bet and passes the bet to the next player.
15 Calling—To match the bet.
16 All-in—When a player bets all of the chips in front of him.

"Damn." Quinton looked at the two pair in his hand. He turned his cards over in feigned disgust. He lost the hand on purpose, but no one knew. Every action and every card was recorded by camera, so he had to play well and play as though he wanted to win, though he didn't. It was masterful.

He continued to do so over the next seven hands until all of his chips were gone.

As people congratulated him for a fine effort, some players gloated about beating him and taking him out of the tournament.

Quinton just smiled as he walked over to the cashier's cage and signed the documents to receive his check.

Confirmation

March 2

THE ANALYST WAS obviously excited and yelled across the lab as she entered the door. "Doctor Spaan. The samples are here!" She hurriedly walked across the room carrying an orange cooler that looked like a picnic style ice chest. A large label was pasted to the outside of the cooler; 'Biological Sample—Keep Refrigerated.'

Imar Spaan was conducting tests on the protein structure of the virus, now nicknamed the 'JC-2 virus,' to see why the virus would have been received into the body and not killed by antibodies as most of the other polyomaviruses are. "Good," he said as he lay some documents aside and met the analyst at the table. "This is a monumental moment. We will either find an abundance of this virus in the samples, thus confirming the presence as an abnormal event, or not." He looked at the analyst. "It is the 'or not' that concerns me," he said as he raised his bushy eyebrows. He opened the ice chest and pulled out a container labeled;

Bonnie Howe—
Cerebrospinal Sample—
February 28, 2015

"Please call Doctor Kruger and Director Pitke down to the lab. They wanted to be present for the test results."

"Yes, Doctor," replied the analyst. The analyst left the room as Imar carried the chest to the isolation chamber of the lab. He put on sterile gloves, mask and gown and entered the lab. He lifted the sealed containers out of the chest and laid them inside a glass box and closed the door. He pressed a few buttons and the box became an airtight chamber. He could hear a vacuum extracting everything in the box except the cylinders. Then, a light cloud filled the chamber. Within a few seconds, the vacuum started again and the cloud was sucked out. The glass box was now a completely sterile environment. Nothing existed in the glass chamber except the sealed samples.

Doctor Kruger walked in as Imar inserted his hands into gloved devices that allowed him to maneuver his hands and fingers inside the box. "I heard the samples arrived," Carl said through the intercom.

"They did," Imar replied, never taking his eyes off the sealed tube he was reaching for. "I plan to run the PCR test and, if it comes back positive, do a visual observation and count of the virus." Imar slowly picked up the canister as he spoke, never missing a beat. He cautiously opened the canister and pulled out a vial of clear liquid. There was a seal on top. Imar picked up a syringe and injected it into the vial, extracting several drops of clear liquid. Then, he picked up another sealed tube and inserted the syringe and squirted a few drops into the new tube. He repeated this step several times until he had six samples extracted from the original tube and placed in new tubes. He performed the same extraction for the 'Stanley Thorne' sample. When he finished, he pulled his hands out of the gloves. "There. Now we can start the analysis," he said as he opened the glass box and removed the vials.

Imar exited the chamber and removed his outer clothing. Carl followed Imar to the PCR machine just as Owen entered the room. "Sorry it took me a few minutes to get here. I heard the samples are in."

"They are," Carl replied. "Imar has isolated the samples and we are ready to run them through the PCR."

Imar slowly inserted a syringe into the vial holding Bonnie's spinal fluid and extracted a few drops. He inserted them into small vials and added the colored dyes and clear liquid. He placed the strip of vials into a PCR machine and pushed a few buttons. "Gentlemen. We will know within the hour if this sample contains genetic markers similar to Mr. Lemolo's, indicating the presence of the virus."

"Well, we should do something while we wait. Any suggestions?" Owen asked.

Carl smiled. "Pray."

—*—

An hour later, Carl, Imar and Owen returned from the cafeteria and walked up to the PCR machine with great anticipation. They were anxious to see the results. Imar pushed some buttons and entered some data on the screen. The printer began whirring as a page slowly ejected. Imar picked up the paper and with a big smile said, "It shows positive."

Owen took the paper from his hand as another page printed. "Well done, Imar. Well done," he said.

Carl took another page from the printer. "This is fabulous," he said. "If Mr. Thorne's sample contains the virus, I believe we will have our smoking gun here."

"What gun?" Imar asked.

Owen and Carl chuckled. "It's an expression, Imar," Owen said. "Conclusive evidence."

"Oh. I see," he said as he removed the vials from the PCR machine and inserted Mac's sample.

"Gentlemen. Do you realize we may well be on the path to finding a cure for Alzheimer's?" Owen asked.

"It is a bit premature, but, yes, I have thought of that possibility," Carl replied. "We still have a lot of work to do, but if this virus is the cause for regeneration of neural synapses, we may have the answer."

Owen was practically giddy. "We could be making history here."

Imar looked at Owen with a slight smile. "Mr. Director, I believe we are."

March 3

EXAMINATION UNDER THE electron microscope confirmed the results of the PCR machine of the presence of the new JC-2 virus in the cerebrospinal fluid of all three patients. The men were ecstatic to find a common denominator in the patients that could well explain why their brains and cognitive abilities were rapidly developing. But there were two questions left unanswered; how did the virus survive the attacks from antibodies and how does it work?

Doctor Spaan continued his research of the properties of the polyomavirus to determine why the antibodies were ineffective in stopping this invader. The regular JC virus was easily stopped by a person's antibodies. This one was not, at least not in these patients. He decided to analyze the protein structure of the virus for any clues. He found none.

Imar began drawing circles on a large whiteboard as he spoke. "It appears the virus is very similar to the JC virus. A person's antibodies should be able to stop the offender with minimal effort. We are unable to answer why the JC-2 virus could survive in these patients," Imar said. He was addressing Carl, Owen, and several key scientists assisting on the project. The group was gathered around a large table in the lab watching Imar write on the whiteboard. "The virus is nearly non-existent in the blood supply, but is rampant in the cerebrospinal fluid. It would appear the brain cells are acting as the host for the virus to replicate."

"I would agree on your assumption that the virus uses the brain cells as the host," Carl said. "We all know that a virus is a single or double strand of DNA. It is not a living organism because it does not contain cells. It requires a host cell, like a bacteria or human cell, to grow. Think of the bacterium or cell as a watermelon, and the virus as a seed. I believe it could be using the human brain cell as that host."

"Can it infect other cells?" Someone asked.

Imar spoke up. "No. I don't believe so." He glanced at Carl for confirmation and continued. "Our tests on other living cells indicate the virus has not attacked them. The antibodies, at this point, are effective with killing the virus in the blood supply."

"But not in the cerebrospinal fluid?" Owen asked.

"Correct," Carl confirmed. "The virus in the blood supply is negligible. The virus in the cerebrospinal fluid is rampant. The environment where it thrives is in the brain and spinal fluid, but we don't know why."

One scientist raised her hand. "How can the virus get into the brain and thrive without passing through the blood supply?" she asked.

"Several ways," Carl replied. "It could be from direct effect where the virus went directly into the brain through an injury or trauma. This is not the case since the patients indicated there was no history of head trauma. Next, through sinus cavities; ears, eyes, nose. If enough of the virus gained access to these areas, it could have successfully worked its way into the cerebral cortex[17]. The third way would be through saturation."

"Saturation?"

"Yes," Carl replied. "That is when the virus is so prevalent in the blood supply that some is able to migrate to the cerebral cortex and take hold before the antibodies destroy it entirely."

"What we don't know," Imar added, "is why it isn't in the blood supply, why it survives in the spinal fluid, how the virus spreads and, more importantly, how it affects the brain function."

"We have analyzed the genetic structure of the JC-2 virus and discovered it contains a similar sequence to the APOE gene located on the nineteenth chromosome." Carl nearly became physically sick when he named the chromosome. The memories of trying to stop the thousands of deaths while working on the 19Q project came over him like a wave. Now, he was dealing with the same chromosome, but a different gene. He took a

---

17 Cerebral Cortex—Outer layer of the brain.

deep breath, composed himself, and assured himself it must be a coincidence. After all, there are multiple genes on each chromosome.

The group became deathly quiet as Carl drew an oblong diagram on the whiteboard. "The APOE gene has been associated with Alzheimer's Disease. It is synthesized principally in the liver, but has also been found in other tissues such as the brain. In the nervous system, neurons preferentially express the receptors for APOE."

"Are you implying that the virus may be able to insert its DNA into the cell and appear like the APOE gene and stimulate neuronal activity?" Owen asked.

"It would appear so," Carl replied. "But we won't know for sure until we have a brain tissue sample to test for confirmation."

The group started murmuring. "This is remarkable, doctor!" one person said. "Astounding!" said another.

"It is, no doubt," Carl said. "But we need that sample to be sure."

Owen stood and pulled out his cell phone. "Carl. We're going to get that sample. You can bank on it." He turned and headed out the door, dialing on his cell phone.

"Bank on it? What does a bank have to do with this discovery?" Imar asked. "Are you going to the bank?"

Carl chuckled as he slapped Imar on the shoulder. "We have a lot of work to do, Imar."

March 5

"THE DOCTOR WILL be right with you," the young lady said.

"That's fine." Mac smiled as he eased into the chair and waited. The room was small with no windows. Everything was white. Mac wondered why they couldn't afford at least a few pictures for someone to peruse while they waited. The magazines were like him; old, used, and deteriorated. His mind was sharp, sharper than ever, but his body continued to slowly break down. Tomorrow was his ninety second birthday, and he was feeling every bit of his age except his mind. The memories of his marriage, the war, his childhood; everything was vivid. At times he was overwhelmed by the reality of the memories and often found himself in tears.

"Good morning, Mr. Thorne," the doctor said as he entered the room carrying a clipboard. "I'm Doctor Lu."

Mac thought the doctor was about the age of his grandson. "Please, call me Mac," he said as he smiled.

"O.K., Mac. I see tomorrow is your birthday. Ninety-two! Wow. That is awesome," the doctor said.

"Yes, it is."

"Any plans for the big day?" the doctor asked.

"Yep. Takin' a nap," Mac said and smiled.

The doctor laughed. "And rightly so." He glanced at the paper on the clipboard and continued. "So, you are here today for a Stereotactic Brain Biopsy."

"I guess," Mac replied.

"This is a requested procedure by Johns Hopkins Alzheimer's Disease Research Center," he said.

"Yeah. They've been doing some tests on me because my memory has improved significantly over the past nine months. They want to know why."

"I can certainly see why they would." The doctor put the clipboard down. "I see you've signed the authorization for us to do this procedure."

"I have."

"You realize, Mac, that this is voluntary. You don't have to do it. There are some risks involved."

"I do." Mac took a deep breath. "You know, doc, when you get to my age, getting up in the morning is a risk." They both chuckled, and Mac continued. "After being in the war and living through that hell, every day I have is a gift from God. I should have died seventy years ago in battle. I'm money ahead. I figure if science can use any part of me to possibly find a cure for that dreaded disease, have at it."

"That's a great attitude to have, Mac. I don't hear that very often from patients."

"Well, you heard it today," Mac said. "You better get going before I change my mind." They both laughed again.

The nurse walked in with a large, round ring with some rods attached to it as the doctor began explaining the procedure. "Mac, we're going to mount this headring to your skull. We will numb the skin and then secure these pins against the skull so the headring doesn't move. Before we do, we will do a CT scan to see exactly where the biopsy will be drawn and then mark the headring. Then we will move you into the operating room where you will receive a light sedation. We will mount the headring, make an incision a few millimeters long in the scalp and drill a small hole into the skull. A thin biopsy needle will be inserted into the brain using the coordinates obtained by the scan and controlled by the headring and computer. The specimen will be extracted and sent to Johns Hopkins. We will monitor you for several hours following the procedure and then you will likely go home." The doctor paused. "There are some risks that I need to make you aware of. You could have an intracranial hemorrhage, seizure, or infection could occur. The probability of any of these happening is minimal, but they are risks. Now, any questions?"

"Nope. Let the games begin," Mac said as he smiled.

—*—

Mac was sitting in the chair much like being in a barbershop. The doctor injected the anesthesia into the scalp. It burned and stung, which caused Mac to remember one of his first haircuts when the barber was drunk. His mother was watching as the barber accidentally snipped his ear. He yelled and cried, but his mother made him stay in the chair and get the haircut. Money was tight back then and she was not about to pay for something and not get it. Today, the barber would be run out of business, maybe charged with some crime.

Oh, how things have changed.

Mac could hear the drill start up as the doctor prepared to enter the skull. He could feel the pressure as the drill bit penetrated the hard bone. There was no pain, but it seemed like there should be. It was like being at the dentist where he could hear the high pitched whine of the drill and knew it should hurt, but it didn't.

"There," the doctor said. "We are through the skull. Now, you'll feel a little pressure as I extract the sample."

Mac was strapped into the chair so his head couldn't move at all. He could feel something in his head, but wasn't sure what it was. It almost tickled. In seconds, the sensation was gone.

"Done. Nice job, Mac," the doctor said as he handed the sample to the nurse.

Mac just sat in the chair remembering the haircut.

March 9

IT WAS LATE in the evening, and Imar Spaan was working tirelessly with several analysts attempting to solve the mystery of why the brains of the patients in the Memory Project were continuing to develop multiple processing capabilities, including memory and cognitive applications. Just the idea that a ninety-two year-old man could recover from Alzheimer's was earthshaking. The potential of discovering a "cure" for Alzheimer's was astounding.

But to Imar, it was the challenge that stirred his blood. Just like on the 19Q project, he knew there was an answer. It was just finding it.

Imar and his team didn't waste any time when the brain tissue samples extracted from Stanley MacKenna Thorne arrived that morning. Imar was confident that the answer was at hand. He immediately split his team into three areas of analyses; genetic, structural, and chemical. Each team of analysts was headed by a scientist specializing in their field of expertise.

Imar supervised structural. He wanted to see how the JC-2 virus infected the brain cells. He knew that a virus would insert itself into a cell and replicate until the cell expanded and died. What he found was truly remarkable.

Imar pulled away from the microscope as if he was shocked, stood to his feet and walked around in a big circle as if confused, and sat back down. A young woman standing behind him watched the strange behavior. Imar carefully leaned into the microscope again, slowly adjusting the focus. Suddenly, he leaned back and smiled broadly and turned to the analyst. "Call Doctors Kruger and Pitke down here right away. Tell them we have the answer."

—*—

The small lab was packed. The three doctors along with several key scientists and some analysts were scattered about the room, standing along walls, around tables and some sitting in chairs. Scientific instruments were strewn about; microscopes, centrifuges, beakers, and a PCR machine for testing DNA. Whiteboards contained scribbles of formulas, diagrams and scientific verbiage that only Imar would understand. It was a place fitting for a discovery of this magnitude.

Imar was standing in front of a small projection screen with an analyst sitting at the monitor. The crowd was silent. All eyes were on Imar and the monitor. "We have confirmed that the JC-2 virus is the catalyst for neurogenesis and cell restoration in the brain," Imar said. "The JC-2 virus contains a DNA and electrical signature almost exactly like that of the APOE gene. It is truly remarkable how closely it resembles the gene." Imar signaled to the analyst who displayed slides of a brain cell and the JC-2 virus in clusters inside the damaged cell as Imar explained the process. "Multiple viruses attach themselves to a damaged neuron and replicate it by interlacing their DNA to the cell. The neuron receives the viruses as it would a normal brain cell with the same DNA signature. The viruses advance tentacle-like arms that allow electrical impulses to pass through the neuron and then from one cell to another and act exactly like a normal, healthy neuron. In essence, the virus is replacing a damaged neuron cell with a functional, simulated live cell."

"This is absolutely phenomenal, Imar!" Carl jumped to his feet. "Do you realize the implications of this discovery?"

"I do," Imar said stoically.

"Are you absolutely certain this is how the virus works?" Owen asked.

"I am," Imar responded. "Without a doubt. We still have more tests to run, but as of this moment, we are certain this is how the virus works. Why, is for another day."

The crowd began murmuring and discussing the discovery. Owen walked to the front of the room to join Imar and Carl. "Ladies and gentlemen. We must keep this confidential. This is a tremendous discovery and if it leaks to anyone, I repeat, *anyone*, it could cause significant, undesirable consequences for all of us." The crowd immediately quieted down as Owen straightened his hair. "We do not need a circus here impeding our progress. What we need is validation of the discovery and a plan to pursue any potential development."

"I agree, Doctor Pitke," Imar said. "We have a better understanding of how the virus works, but we do not know how it enters the body, why the immune system allows it to progress when the typical JC virus is easily controlled and what the virus uses to multiply while in the body."

"So noted, doctor," Carl said. He turned to the audience. "I think those are our marching orders. Let's meet back here early tomorrow and get these answers."

The group slowly disbanded and filtered out of the room. Imar, Carl and Owen stayed behind. Owen was first to speak after the room emptied. "Do you realize the implications of this discovery?" Owen was nearly giddy with excitement as he spoke. "If this virus creates a sort of simulated neurogenesis in the brain, it could well expand to multiple applications: brain injuries, encephalitis victims, brain damage from oxygen deprivation, paralysis regeneration and who knows what else." He sounded like a kid reciting his Christmas list.

"Hold on, Owen," Carl said. "This is huge, but we have to take this one step at a time. It works on damaged cells; not dead ones."

"I concur," Imar said. "We mustn't get the cart out of the barn too quickly."

Both men looked at Imar, and laughed. "You mean get the cart before the horse, or close the barn door once the horse is out. Not both," Carl corrected.

"Is that not what I said, but more succinctly?" Imar asked.

The men were puzzled. To some extent, Imar was correct. They both laughed again. "Well, I guess you did, Imar," Owen said.

Carl brought them back to reality. "However we say it, let's take one step at a time. We have a starting point."

The men discussed the situation a little longer and agreed that they would focus on the task at hand and solve the remaining mysteries of the JC-2 virus. Each man, however, knew very well that this was a discovery

that could eventually lead to a breakthrough synonymous with pasteuriza-
tion or the polio vaccine. Their names would go down in history. They each
would undoubtedly be remembered as one of the greatest scientists of all
time.

March 18

THE PAST TWO weeks of research were wearing on Imar and Carl. Both men worked long hours trying to solve the final mysteries of the JC-2 virus with their research teams and the best equipment available. Owen Pitke and the Johns Hopkins board did not hold back anything for the research teams. They had access to anything and everything: the best facilities, equipment, people, money, whatever it took. The intent of the board was clear to Owen; find out how this virus cures Alzheimer's and bring it to market. Period. It would be the greatest discovery of the new millennium, possibly of the past hundred years, and Johns Hopkins wanted every cent they could make from it to advance future research programs: cancer, diabetes, sickle cell, everything. This was going to be the cash cow for the university and they couldn't wait. Owen couldn't wait. The world couldn't wait.

Finally, the teams had all of the answers. They knew the inner-workings of this new virus. They knew the JC-2 virus entered the body through a strain of bacteria common to sewage treatment called the acine-tobacter. The bacterium was the host for the virus outside the body and could survive for weeks on a dry surface, but the virus could not. That meant the virus had to enter the body within days of infecting the bacteria, drawing the assumption that the virus entered the body through a water or liquid source. The body needed a low immune system for the bacteria to spread since the body's antibodies could easily kill both the bacteria and the virus. They confirmed this by researching the medical records of the patients and interviews and confirmed all three had compromised immune systems about the same time in 2014. Their previous research missed the fact that Quinton Lemolo had a serious cold. Their compromised immune systems allowed the bacteria infected with the virus to enter their bodies and proliferate.

Once inside the body, the virus and bacteria would localize in the brain through something similar to meningitis. The bacteria allowed the virus to continue to spread and attack damaged neurons in large quantities. Once a cluster of viruses entered a damaged cell, it would inject its DNA causing the cell to look like it was a living cell again. In reality, the virus was creating healthy looking neuron impersonators by using multiple viruses as DNA markers. It was like filling a mold with plaster to look just like the original. What baffled the scientists was that the virus didn't kill the cells like it did the bacteria. Instead, clusters would form and continue to live in the cell. The virus would continue to replicate and once the cell was full, extra viruses would escape through small cell fissures and infect surrounding damaged cells. The bacterium was the host for replicating and introducing the virus and the brain cell was the host for maintaining its existence and expansion.

The DNA of the virus contained markers so similar to the APOE gene that live neurons that attached to the infected one saw no difference and allowed electrical impulses and nutrients to pass through to the cell. The combination provided a type of resuscitation to the cell, bringing it back to life as water would on a dead plant. The phenomenon passed from one cell to another to another until the cells grew together and repaired the tangles of the brain, allowing electrical impulses to pass properly through the synapses of the brain. It was similar to a deep brain electrical stimulus[18]. A disease that bridged the gap of Alzheimer's. It was truly a remarkable breakthrough.

And Imar, Carl and Owen knew it.

—*—

Owen was sitting at his desk talking on the phone. His voice was low so the others couldn't hear him. Carl and Imar were sitting across the room

---

18 Deep Brain Electrical Stimulus—Brain pacemaker that sends sporadic electrical shocks to the memory area of the brain to stimulate memory retrieval.

admiring the sunset out of the expansive windows of the director's office. "Do you think they will authorize the next phase?" Imar asked.

Carl turned and smiled. "Without a doubt."

Owen hung up the phone and jumped to his feet. "Gentlemen, we are in!" he said with great excitement.

Carl smiled at Imar. "See?"

"The board reviewed all of the materials, tests, results and reports and just approved full scale testing to develop an inoculation of the JC-2 virus!"

"That's fabulous, Owen," Imar said.

"Fabulous? Imar, it is much more than fabulous. It is history my dear friend. History!" Owen reached out and grabbed Imar's hand and shook it heartily. "We are proceeding with the development of a cure for Alzheimer's, at a minimum. Alzheimer's!" Owen laughed again and shook Carl's hand. "Can you believe it? Our names will go down in history with the likes of Pasteur, Jonas Salk, …"

"Owen. That's great, but this is not about our names and history," Carl said.

"Sure it is, Carl," Owen retorted as he straightened his hair.

"Maybe for you, but not for me." Carl took a deep breath and continued. "If we find a cure and that happens, great. But first, we need to find the cure. We have tests that indicate it might be this virus, but we have a long way to go."

"Carl. You're a wet towel; noble, but a wet towel," Owen said.

"A towel?" Imar asked.

The two men ignored Imar's question. "I know there is still a long way to go, but this is the first step toward something historical. Don't squelch the moment." Owen said. "That call was a green light for full scale testing; no holds barred, no limits."

"What?" Imar asked.

"We can do anything we see fit to bring this to a viable product for market." Owen placed his hand on Imar's shoulder. "In summary, my good man, we can make the cure."

Imar smiled. "It would be nice to follow this through to the end, don't you think, Carl?"

Carl turned and looked out the window. He had a sense of conflict arising. He wanted to see a cure, sure, as did everyone. He wasn't sure why he felt this way. Being identified as one of the creators of the greatest cures of all time was terrifying to him. He had seen famous people go down in flames. "Sure, Imar. It would."

March 30

OWEN PITKE DIDN'T waste any time. Within two weeks he was able to procure the funds from the board of the University and establish a separate account for the management of the Memory Project. Even the board adopted the name given by the doctors. It fit. It was easy to remember. It represented the University well. They were geniuses.

The project was focused. Create an inoculation of the virus to cure Alzheimer's. Other dementias may well be cured, but the single goal at present was to create a cure for Alzheimer's and bring it to market as fast as possible. The virus was in the environment, somewhere, which meant someone else could happen upon it, discover its function, and create a cure before Carl's team. Owen wouldn't think of it; he wouldn't have that. He was determined to drive the team as hard as possible to find a way to bring this product to market as fast as possible.

He immediately started the conversion of a large research wing after he got the nod from the board. The existing wing was an established biohazard three facility often used to analyze and isolate various viruses. That gave Owen an edge. He had crews work around the clock to install the most current security and decontamination equipment available and upgrade a section of the wing to a biohazard level four facility. The new wing would be specifically used to create this new medicine and nothing else. He knew they would need a completely secure and sterile environment to accommodate the extensive testing on animals and humans. The product would undergo extreme scrutiny before it was considered applicable to Phase One clinical testing. Time was of the essence, and Owen knew it.

In just two weeks, the new wing consisted of several laboratories fully staffed with the best equipment and scientists JHU could muster. The access was completely secure with one entrance staffed by several security guards and access protocols. Nothing could go in or out of the fully contained research area without extreme decontamination procedures. Once inside, the wing was divided into four areas of focus: inoculation, chemical composition, application and culture. The latter contained level four biohazard protocol. It was imperative that the virus cultures were not contaminated in any way.

"Good morning, Doctor Spaan." The guard was pleasant as he stood and checked Imar's identification. Two armed security guards were scattered in the reception area; one at the entry hall and one by the entry door to the test facility. "Thank you, sir. Please proceed, sir." Imar approached the guard by the door and leaned forward as the retinal scanner scanned his eye. "Imar Spaan, one three seven six."

"Identification confirmed," the automated voice replied. The door buzzed and opened. Imar passed through as the door closed behind him and clicked. The entryway was circular with hallways branching off in three directions. The room was bleak, white, sterile. Windows ran the length of the hallways on both sides starting a few feet above the floor and rising four feet. Imar could see the people behind the glass working on a variety of tasks: centrifuges, animals, test tubes, beakers, and so on. Through a glass wall across the room Imar could see several yellow biohazard suits and masks hanging on the wall. In the opposite direction he saw several contractors carrying light beams and countertops. It was a circus of activity trying to create the cure and continue to update the facility. Directly in front of Imar was another door with a sign reading "Lab A." An armed guard was standing in front of the door. Imar walked over to another scanner and placed his palm on the glass screen.

"Imar Spaan, one three seven six."

"Please proceed Mr. Spaan," the computer said as the door clicked open. Imar walked through and the door closed. Imar walked to the center of a small entry room and placed his feet in two footprints outlined on the floor. "Please extend your arms and stand still," the computer voice said. Imar followed directions and a great wind entered the room for a few seconds ruffling his already wild hair. The wind stopped. "Thank you. Please proceed." The door behind Imar clicked open. As Imar walked into the Memory Project culture lab, the door closed behind him with a click.

"Imar!" Carl walked briskly over to Imar to greet him. "We are making great progress here," he said as he pointed to several chambers behind glass walls where cultures of live virus could be seen growing inside glassed walls. Two people were inside wearing biohazard suits and masks working on some cultures on the counter. "We have several live samples ready for introduction into some test subjects."

"People?" Imar asked.

Carl laughed. "Of course not. Animals. We plan to start with Wistar rats and move to monkeys if things go well."

The white Wistar rat is the most common species of animals used to test a variety of clinical applications in laboratories. Typically, the subjects are exposed to various levels of vaccines containing dead viruses or bacteria to see if their immune systems can create antibodies to fend off healthy viruses or bacteria. This case was different. The Memory Project team hoped the live viruses embedded in the acinetobacter bacteria would be received by the subjects and move to the brain to grow. They didn't want the antibodies to build and kill the virus; they wanted the virus to sneak past the antibodies.

"Of course," Imar said, snickering at his silly question. "When will we start the introduction?"

"This afternoon. We will have samples ready for injection. It is much easier than preparing a vaccine since we are introducing live viruses into the subjects. I know the test subjects are already prepped."

Imar stared at the culture dishes in the case. "Then we are ready to begin."

"We are, Imar. We are."

—*—

The rats were separated into four cages with nameplates on each cage. Each rat was named, as was always the case in the JHU test facility. Leroy, Gabby, Casey and Alejandro; two female, two male. An analyst wearing a biohazard suit opened the cage and reached in to grab Gabby with her gloved hands. She held the rat on a small tray as another suited analyst raised the scruff of the rat's neck and inserted a needle with a syringe. The analyst slowly injected the rat with the virus. They placed the rat back in the cage and jotted some notes on a pad. The analyst who injected the rat entered some information on a screen. The pair continued to inject the rats, one by one, until all four rats were infected with the virus.

Leroy

April 24

NO ONE KNEW how the test subjects would receive the bacteria or the virus. The bacterium was common in the world, and most rats and humans had it. So the hope was the body would look past the fact that the bacteria were infected with the JC-2 virus and just allow it to enter and stay in the body.

Gabby died within four days of the injection from a bacterial infection. An autopsy showed Gabby had deficient antibodies to fend off the infection. The other three rats did just fine.

Casey and Alejandro showed no sign of improvement. Autopsies revealed the virus did not take hold. The immune systems of the rats effectively killed the virus.

Leroy instantly became ill and quite lethargic for two days. Everyone thought he was going to die, but he pulled through. For several weeks, he functioned very much like a normal rat. He watched the other rats in cages nearby wishing he could play with them. He was resigned to play with his abundant toys and wheel.

Then came testing day.

Every three days Leroy was removed from his cage and tested for basic memory development. A maze was created each test day that had a piece of cheese at the end. Leroy was placed at one end of the maze and allowed to wander through the twists and turns until he arrived at the cheese. He was allowed to eat a little, and returned to the start of the maze. The analysts would time the second and third pass through the maze to see if Leroy remembered what turns to take and how fast he could finish the maze. The maze was changed each test day to assure Leroy was remembering the path of that day and not historically.

"O.K., Leroy," the analyst said. "Here ya go." The girl placed the rat at the beginning of the maze and tapped a timer while another analyst checked the video camera and wrote down a couple of notes. They watched the rat come to a wall, turn left, then right, and continue to struggle through the maze until it arrived at the cheese. "Eighty-six seconds. Not bad for a rat," the girl said.

The young male analyst checked his notes. "We had thirty-three decisions and fourteen corrections," he said.

"O.K., Leroy. Round two." The young girl placed Leroy at the start of the maze and hit the timer. Leroy took off like a lightning bolt, scampering through the twists and turns.

"What the ...?" The young man watched intently without taking a note. The young girl stared at the rat running through the maze. "Look at him go! Look at that!" It was like she was watching a race.

"Go, Leroy!" the young man shouted. Carl heard the two yelling as they watched Leroy dash through the maze. Suddenly the girl yelled "Fifteen seconds!"

"Woo-hoo!" The young man shouted. "No errors!"

Carl walked up to them. "What's all the excitement?" he asked.

"It's Leroy, Doctor Kruger. Watch!" She placed the rat at the start of the maze and hit the timer. Leroy bolted.

Carl stared as the rat ran at breakneck speed to get to the cheese. "Are you kidding me?"

Ding!

"Thirteen seconds! He beat his time!" the girl shouted.

"O.K. That is impressive. Double the maze," Carl said.

He watched as the pair pulled another maze over to join it to the existing one. They moved some walls around and connected the two, shifted walls in the first maze to create a new path, and ran the test again.

Then they connected a third maze and ran the test.

And a fourth maze.

After the fourth maze, Carl pulled out his cell phone. "Owen. You need to come down to Bravo lab and see this. Yes, right now. Bring Imar."

—*—

The group had seats around the large area converted into an Olympic type rat racetrack. Some people were standing, others sitting on the floor or on counters. Everyone had a view. No one wanted to miss this.

In the middle of the room were six interconnected mazes. The start was labeled and the finish line had a piece of cheese on a small tray. Leroy was sitting behind a wall pacing back and forth, eager to run the race to the cheese. People were murmuring, wondering if he could do it and how long it would take. Several people placed side bets. Some were laughing. Everyone was watching.

Carl, Owen and Imar were standing center court watching the event unfold.

"How many decisions are in this configuration?" Owen asked.

The male analysts jotted some numbers down. "Two hundred and seven, sir."

"This is impossible," Imar said. "No living thing can remember that many decisions on the first try."

"We shall see, Imar," Carl said. "O.K. Are you ready?"

"Yes, sir," the young man said. The girl started the timer and opened the gate. The crowd watched as Leroy dashed from side to side, finding the route that led him through the maze. Several paths went for feet before he had to turn around and start all over at the next gate. Everyone watched intently. The crown cheered as Leroy made the right decisions. After sixteen minutes, it appeared he would finally make it to the end and get the cheese.

"He must be exhausted," someone said from the crowd.

The female analyst responded. "This is the seventh maze today. I bet he is tired. He drank quite a bit of water," she said.

"That's O.K.," Owen said. "Let's see if he can remember any of the moves anyway."

The girl lifted Leroy to her face and whispered, "O.K., Leroy. Show 'em what you've got." Everyone closed in as the girl placed Leroy behind a screen. The girl hit the timer and opened the screen.

Leroy was off.

"Look at that!"

"Wow!"

"Look at that rat go!"

"White lightning!"

"Amazing!"

People were yelling, pointing, clapping, whistling, watching, slapping each other. Carl watched intently. Owen smiled. Imar shook his wild hair in disbelief as Leroy finished the maze.

"Eighty-four seconds!" the girl said. "Eighty-four!"

"No errors," Carl said. "None. Not one."

Owen was smiling broadly. He looked like it was Christmas morning and he was ten years old. "Success! Gentlemen. We have success!!"

Imar looked at Carl. "Incredible," he said as the audience applauded loudly.

Origin

May 11

LEROY, THE RAT, was already a legend at the Johns Hopkins Alzheimer's Disease Research Center. Over the next few days the team ran test after test to challenge his memory retention and cognitive abilities. Leroy could learn to ring a bell for food on the first try. He was able to figure out that different buttons meant different foods or failures, and he quickly knew which buttons to push for the best treat. He pushed buttons, climbed confusing configurations to reach his goal, and performed tricks for rewards. He was a star. No rat before him ever accomplished so much. He was talked about by everyone, everywhere, but only in the center. No one on the outside knew he existed. The team couldn't run the risk of their experiments leaking out. They had to be first to develop and market this new wonder drug.

Leroy's use was complete. It was time for him to pay the ultimate price in research. It was time for him to die.

Carl and Imar performed the autopsy on Leroy. They dissected the little star and removed his brain for closer examination. They confirmed the virus entered the brain and began multiplying at a rapid rate, infecting damaged cells and bringing them back to 'life.' Leroy's brain development was astounding. In just six weeks they determined his brain grew ten percent by volume. They estimated he had doubled his processing and cognitive abilities, but were unable to confirm it without a PET scan of the living subject.

The other rats in the research center who successfully received the virus also demonstrated improved memories within a few weeks. They were prodded, tested, and re-tested just like Leroy. All of them excelled, but Leroy was the first. Surprisingly, only two additional rats died. The rest of the rats received the virus without side effects, all forty-eight of them. Imar figured out how to encase the bacterium in a protein shell to cloak the bacterium into an acceptable invader of the body. The antibodies saw the bacterium as a type of blood cell derivative and allowed it to pass through the system directly to the brain where it took hold in the damaged and dying cells and proliferated. The other bacteria that stayed in the bloodstream eventually died as the shell collapsed. The only explanation for the virus to be attracted to the brain cells was it had a similar electrical impulse and the brain cells had just the right amount of porosity. It was only conjecture since the virus appeared to be effective in any mammalian brain. No one could prove it at this point. It didn't matter anyway. They had a method to effectively infect the host with the virus and enable it to settle in the brain to grow, and that was all that mattered.

Carl and Imar were discussing how to proceed with the next phase of testing while simultaneously testing the rats when Owen walked into the lab. "Where are we, Carl?" he asked.

"Imar and I are discussing simultaneous experiments across varying subjects. We think we should move to the cynomolgus monkeys from the Tinjil Island breeding facility."

"Tinjil?" Owen asked. "Where is Tinjil?"

"Indonesia," Imar replied.

"That's halfway around the world!"

"I know," Carl replied.

Owen looked at Imar and straightened his hair. "Why can't we use local monkeys?"

Imar hesitated for a moment. "We can, but it will not assure us that we have a viable product for testing on humans. The Tinjil monkeys have been bred to provide subjects free of specific diseases common to their species, including Herpes B, simian immunodeficiency virus, and simian retroviruses."

Owen was stressed and straightened his hair again. "Will they have an adequate supply?"

"I took the liberty of contacting the facility director. We will start receiving the monkeys in a week if you give the green light," Carl assured Owen. "We can start with six monkeys for staggered tests. That will assure us of two active test monkeys at all times. The facility has several hundred and the director assured me they can meet our demand."

"We will also conduct simultaneous testing on more rats and some ferrets," Imar said. "We want to cover all of our vases."

"You mean 'bases,' Imar. Cover your bases," Carl said.

"Yes, I suppose."

"Look. As long as we don't lose any more time on this. The board and I are concerned about someone beating us to the cure. We know of two more cases in Sacramento, so we ..."

"Sacramento?" Carl asked. "Did you say 'Sacramento'?"

"Yeah. Sacramento," Owen replied. "What's the big deal?"

"Then, where is Elk Grove?" Imar asked.

"It's just south of Sacramento, about twenty-five miles." Owen could see the concern on the faces of both men. "What's going on here?"

Carl threw the papers he was holding onto the desk in disgust. "Why didn't you tell me this started there?"

"I didn't think it mattered!"

"It does!" Carl yelled.

"Why? Will you tell me what you're all upset about?" Owen pleaded.

Carl motioned to the chairs by the table. "I think we better sit down. This is going to take awhile."

—*—

Owen stared out the window. He was stunned. He just spent thirty minutes listening to Carl and Imar tell about the 19Q project from four years back. He listened to them describe the agony of attempting to solve the mystery of why so many people were dying from Maskill's Seizures. He listened to them tell the tale about the deaths of the owners and a co-worker, the ensuing FBI investigation, the creation of the chemical element that caused the mutated wheat to kill so many people, and the subsequent bombing of the town of Quincy, Washington. He felt like he was listening to a book on tape; a movie being played back.

"I recall the Maskill's fiasco," Owen said. "I had no idea you were involved. That is quite a tale you have there, gentlemen," Owen said as he smiled and straightened his hair.

Carl became incensed. "This is not a joke. It is not a tale." His voice rumbled through the room. "We watched more than sixty thousand people die while we struggled to find a way to stop it." Carl always felt like he didn't do enough to stop the deaths. People considered him to be a hero. He didn't.

"So, you think there is a connection to this virus?"

"Yes, I do," Carl said. "I can feel it in my gut. Something's wrong."

"It is quite a coincidence that this virus originates from the same vicinity as the company that created the chemical element," Imar said. "And we know that chemical caused the mutation in the wheat resulting in the deaths."

Owen shrugged his shoulders. "So?"

"So? What do you mean, 'so'?" Carl stood to his feet positioning himself as though he was ready to fight. He could feel his blood pressure rise. He felt like Owen was demeaning everything he just shared. "This is no joke, Owen."

"Settle down, Carl. I don't mean to be disrespectful. I'm a realist," he said and straightened his hair again.

"Realist?" Imar asked.

"Yeah. So what if this chemical agent is involved. It doesn't change the fact that the virus is doing good by curing people like Mac with Alzheimer's, does it?"

"I ... I suppose not," Imar said.

"It can't be right, Owen. How can something so deadly become something so helpful? I don't trust it." Carl walked over to the window. "We watched a lot of people die, families being torn apart. It was a nightmare." Carl paused, remembering the people suffering with losing their children, parents and spouses. "Something's wrong here."

"Well, Carl. I'll believe something's wrong when I see it. Until then, all I see is an amazingly good cure for a terrible, debilitating disease," Owen said.

Carl turned to Owen. "All you see are dollar signs from the development and sales of this cure."

"Back off, Carl. I'm still your boss."

"Carl," Imar interrupted, attempting to break the tension. "We can at least continue the testing, thoroughly, before we recommend to proceed with development of the virus. We have some control and assurance it will be O.K. before it is released."

"And the FDA will play their part as well," Owen assured him. "We have all the checks and balances we need to make sure it works, Carl, so no need to fret about it."

Carl walked out of the room.

Owen turned to Imar and pointed his finger at him. "You need to make sure he's on board. And order the damn monkeys." He picked up his papers and left.

Imar gazed out the window.

—*—

The drive home was arduous. Carl thought about the 19Q project and the struggles they had trying to identify the catalyst for the deaths. The only good thing that came out of that project was meeting Alice and falling in love. Other than that, Carl felt like the project was a failure, a near disaster. If they had not received the letter from the partner telling about the wheat mutation, they may very well have never figured it out. He didn't want to go through that again. No one in their right mind did. If only Owen knew how bad it was. There must be some way to get him to see this is wrong.

Carl instinctively pulled up to the house without realizing he was even home. He was deep in thought as he walked in and was greeted by Alice. "Hey, handsome." She struggled across the room on her crutches and grabbed his arm to pull him down for a kiss. Carl didn't look at her until she grabbed his arm, and she knew something was wrong. "Where are you?"

"Huh?"

"I can see you are somewhere else in thought, so where are you?" she asked.

"Nineteen Q," he said solemnly.

Alice was shocked. "Why are you thinking of that?"

"Because we discovered that the patients exposed to this virus are from the Sacramento area."

"Are you certain?"

"Yes. The town they are from is a few miles south of Sacramento. That is too much of a coincidence for me."

Carl and Alice walked to the sofa and sat down. "Have you done any tests to confirm the presence of the chemical element?"

"No. I was thinking of that on the way home. I need to have someone do the molecular analysis to confirm or deny its existence in the virus." Carl paused. "What if it's there?"

"Then we'll cross that bridge when we come to it."

"How can you say that? You saw the devastation that stuff did with Maskill's."

"Carl, I can say that because I want to believe it is a possible cure. You know Alzheimer's runs on my dad's side of the family. He died from it. I'm scared I could get it."

Carl hugged Alice. "Hey, baby. You won't get it."

Alice became irate. "How do you know? How can you say that? You don't know what I might get." Alice grabbed her crutches and stood.

"Hey, settle down." Carl stopped her. "I just want to protect you and take care of you."

"Then don't stop something that could be good because you're afraid of something that happened in the past."

"And don't be blind to something because of your fear of the future," Carl shot back.

"Arrgghh." Alice slapped Carl with her crutch and hobbled off.

"Why did I say that?" Carl mumbled. He lightly hit the side of his head saying, "Stupid, stupid, stupid."

May 15–18

CARL WAS SITTING at his desk reviewing email when he saw one from Norman Raynould, the molecular physicist that worked on the 19Q project with Carl.

> To: Carl Kruger
> From: Norman Raynould
> Subject: Molecular breakdown of JC-2 virus
>
> ---
>
> Carl;
>
> It's there. Sorry for the bad news. Let me know if you need me there.
>
> Norm

Carl stared at the message. His fear was realized. The chemical element Doctor Everett Maskill created that killed thousands of people four years ago was in the JC-2 virus.

—*—

Jamille Larson was an ambitious young man. At forty-one he was one of the youngest Supervisory Special Agents with the Bureau. The work he did on the 19Q project a few years ago catapulted him to the new position. He always felt he had to work harder than the other agents to prove himself.

Maybe it was his history of struggling through school as a black child in the inner city. Maybe it was his mother always telling him to work harder, be smarter and get an education. Maybe it was his dad telling him he'd never make it. Whatever the motivation, Jamille surprised everyone, including himself, when he received his degree in criminal justice from Stanford University and went on to join the Federal Bureau of Investigation. He had a knack for investigation. He was naturally a curious man who needed answers. Now, he was one of the agents in charge of the field operations in Sacramento, California. He loved his job. Life was good. Very good, even though he had to deal with the worst of the worst in our society. It gave him a great sense of contributing to the good.

Jamille was looking over some reports when the phone rang. "Agent Larson here."

"Agent Larson. Doctor Carl Kruger here." Jamille smiled widely as he listened to the voice from the past. "I bet you never expected a call from me," Carl continued.

"Carl Kruger. Man, it's great to hear your voice." Jamille thought of the work they did together, the friendship they formed during their work on 19Q. "How are you? Still married to Alice?" Jamille recalled their wedding and the prank he and Bob Struthers pulled when they wrote 'Help Me' on the bottom of Carl's shoes so that when he kneeled to take communion, the words appeared and the audience started to chuckle.

"Jamille, I'm doing great. Yes, still married and have adopted a young boy. Joey. He's six and a handful. And you?"

"That's great! Still the same here. Just running the field office with a partner. Keeping pretty busy." Jamille paused. "I take it this is not a friend-ship call, Carl."

"It is, sort of." Carl took a deep breath. "I need a favor from a friend."

"As long as it's not illegal, you've got it. What is it?"

"I have a situation here where the Maskill chemical element has resur-faced." The phone was silent. Carl waited for Jamille to respond. "Jamille. Did you hear what I said?"

Jamille leaned forward in his chair. "I did. I ... I just can't believe it. You mean the one Doctor Maskill created that caused all of the havoc with 19Q is back?"

"It is. But it's different this time."

"What do you mean 'different'?"

"It's curing people … not killing them."

"Well, I'd say that's different for sure."

"Jamille. We have a virus that has infected a few people in the Elk Grove area near you. The virus is curing some significant diseases."

"Curing? Like what?"

Carl hesitated. Did he dare tell Jamille what disease they were trying to cure, or would the news leak out? He knew he could trust him, but what if someone else discovered what Jamille was working on and leaked it? What if another company started research based on something that leaked from …

"Carl. What is it?"

"Jamille, I'm not sure I can tell you. I trust you implicitly, but if this leaks out, I could have a very real problem here."

"At Johns Hopkins?"

"Yeah."

"O.K. So, what can I do to help?"

"I need to know if there is a link between Biotrogen, the shell company, and Elk Grove. That's all."

Jamille thought for a few seconds. "So, you just need to know if Elk Grove is mentioned anywhere in the records of both companies, and why? Is that it?"

"Yes. If there is a connection, it might show how this virus came into contact with the chemical and eventually into these people."

"But you said it was a cure; a good thing."

"It is, but I have my reservations. After 19Q, I just can't see this stuff doing anything good. I need to know."

"Okay. I can do that." Jamille started to write down some notes. "Can you tell me the names of the people or where …"

"No. Jamille, I really can't tell you anything more. I just need to know if there is a connection and why."

Jamille got irritated that Carl was only willing to work one way and trying to limit his understanding of the situation. Jamille was used to being in charge and he needed to know what the situation was if he was going to be able to help his friend. Besides, his interest was almost out of control and he wanted to know; he had to know. "Look, Carl. I need a little more to go on. I want to make sure I'm not doing anything wrong on my end. You're asking me to go into a closed case and research some records. I should at least know why in blazes I'm doing that."

Carl almost wished he hadn't called Jamille. "Jamille. I need your word this is not going any farther than you."

"You've got it. Now, what are we doing here?"

Carl took a deep breath. "We're looking at a cure for Alzheimer's."

Jamille was speechless.

"Jamille? Jamille?"

"Uh, yeah. Carl. Wow. I mean … This is … wow. Alzheimer's?"

"Yeah. It could change the world."

"O.K. Uh … I'll get right to this. I … uh … I can do a quick word search through the company files we extracted from Biotrogen and just see if Elk Grove is recorded anywhere. I can search both the real and shell companies pretty quickly."

"How quickly?" Carl asked.

"Two days."

"Monday?"

"Yeah. I didn't have any plans this weekend. I should have an answer by then. If there are multiple references, maybe longer. I'll document what I can."

"Jamille. I can't thank you enough. Let me know as soon as you find something."

"Sure will, Carl. Good to hear from you, all things considering."

"And you."

Jamille hung up the phone, sat back in his chair and quietly mumbled, "Alzheimer's."

May 18

ALICE BELIEVED THEY should keep the landline phone because of the difficulty her parents had when they visited. They couldn't use a cell phone because it was too complicated. They had to use a push button phone or they were lost. With her dad passing, she wanted to make sure her mom never had to deal with a phone number change and could always access a phone if she came to visit. Times had changed, but not for Alice's mom or Carl's parents. Besides, it was good to have a house phone for those times their cell phones needed charging or were misplaced, as was Carl's this morning when he left it at the office the night before.

Alice was picking up some of Joey's clothes when the house phone rang. "Hello?"

"Is this Alice Barker-Kruger?"

"Jamille? Jamille Larson?"

"The one and only."

"What a surprise!"

"Yes, it is Alice. I was calling Carl, but your voice is so much nicer to hear than that bear."

Alice blushed. "He is sometimes. Let me get him." Alice turned and yelled up the stairs. "Carl. Jamille Larson is on the phone." She turned back to the call. "He'll be right here, Jamille. It's good to hear from you. How have you been?"

"Great. Busy, but great."

As Jamille answered, Alice could hear the 'click' as Carl picked up the phone upstairs. "I've got it, babe," he said.

"O.K. Good to hear from you, Jamille."

"You too, Alice."

Alice wasn't accustomed to eavesdropping, but she felt there was something unusual about a call from an FBI friend, and she was curious. She covered the mouthpiece of the receiver as she tapped the phone hook. Carl heard the 'click' and thought she had hung up. "What did you find out, Jamille?"

"I'm sending you an email today of the list of references. There's only a few. Some are addresses of people they had contact with. There were just a few basic companies; a bakery, flower shop, stuff like that. One in particular caught my eye. It was a storage company used under the shell company books."

"Storage?"

"Yeah. Looks like they paid several years in advance for a unit. The unit expired last year."

"Last year. That was when this surfaced." Carl got excited as he spoke. "Listen, Jamille. We need to find out what happened to that storage unit. Is there any way …"

Jamille interrupted him. "Carl! Do you know what you're asking? You're asking me to conduct a covert investigation of a closed case."

"No, I'm not. I'm asking you to just look into it … a little. Maybe call the storage company and see if they can share anything."

"I can't get any information unless I have an open case."

"I don't want to use it. I just want to know if Biotrogen is involved with this new virus."

"I understand." Carl could hear Jamille sigh. Alice could, too. "Look. I'll ask some questions and do a quasi-covert call-around investigation type thing. This is just a superficial inquiry. I hope I don't find anything significant here or that will create some problems."

"We don't want any problems, Jamille. None at all."

"Agreed. O.K., I'll see what shows up."

"Thanks, Jamille." Carl could hear Jamille hang up. He thought he heard a second click just before he put the receiver down.

Strange.

Albert and Art

June 1

MONDAY MORNING AND Carl was just getting ready for work. Joey was already on the bus. Alice was getting ready to go to her mom's house to run her on a few errands. Ever since her dad died, her mother became more dependent on Alice and Carl to help hold things together. Her mom had to stop driving last year, with great resistance, because Alice contacted DMV and recommended she have a driving test before they renewed her license. Now, she is suffering the lack of independence that comes along with owning and driving a car. She is dependent on people, taxis, busses and anyone else that can give the elderly lady a lift. She hated to ask Alice for rides, but Alice brought this on herself. Now, she can just drive her around town.

Today, it was to the salon to get her hair done. A woman can lose her memory, but she doesn't lose her desire to look beautiful. The hair must be done.

Carl was just getting his jacket when his cell phone rang. "Hello?"

"Carl. I think we may have a problem."

"Imar. What is it?" Carl sat his coat back on the chair.

"Little Bonnie Howe just died."

"What? Bonnie?"

Alice could hear Carl talking about Bonnie and came closer to hear. "Yes. She went into the hospital Saturday and died early this morning. I got the call from her doctor. He knew she was in the test group."

"Oh, that's terrible. What did she die from?"

"They think it was the leukemia, maybe the chemo, not sure. I think they should do an autopsy to be sure, don't you?"

Carl thought. "Yes, I do. Will they do it? Do you know?"

"No, I don't. I'll call Owen for a meeting this morning to see how to approach this."

"Good idea. I think we want to walk on eggs here," Carl said.

"Eggs?"

"Yes. Be careful, go slowly."

"Oh. Yes. Eggs. O.K., then. I'll see you when you get in."

Carl closed the phone and looked at Alice. "Bonnie died?" she asked.

"Yes. I hope it wasn't from this virus. I just have a feeling," he said.

"Well, let's just take one step at a time. I heard Imar say something about an autopsy. I think that is a good idea, if you can get them to do it."

"If," Carl agreed. "I gotta go, hon." Carl kissed Alice on the cheek. "Love you."

"Love you more," she said as he walked out the door.

—*—

Imar, Owen and Carl were meeting in the lab before anyone showed up for work that morning. Imar set the meeting up and he always felt more comfortable in the lab than an office. "Carl, you're really starting to worry me." Owen straightened his hair and continued. "Ever since you found out this virus was associated to some chemical element you just won't let this die."

"Poor choice of words, Owen," Imar said.

"Both of you …" Owen's face became red and flushed. He pointed to them and continued. "… are really pushing it here. I'm beginning to think you may jeopardize this whole project."

"And just how would that be, Owen?" Carl asked.

"I don't know. I don't know if I have to replace you or watch you or what."

"Listen, Owen." Carl leaned forward resenting the fact that Owen thought he would do anything to jeopardize the project. "I'm a professional. I may not agree with something, and, if I feel strongly enough, I'll leave, but I won't sabotage any work we are doing."

"Are you going to leave then?" Owen asked.

Carl thought for a second, longer than Owen would have liked. "No. Not if the autopsy on Bonnie comes back negative on the virus as the cause of death. If it's positive, then either we pull the plug on development, or I'll leave."

"Okay. Fair enough. Let's get this autopsy." Owen stood to his feet and left.

—*—

It was evening and everyone left the lab except the guard. The area was secure and under tight surveillance with cameras and such. "Good evening, director," The guard said.

"Yes, it is," Owen said as he punched in the code, looked into the scanner and repeated his code for voice recognition.

"Working late?" The guard asked.

"No. I think I left my cell phone in there."

The door opened and Owen walked into the next secure area with the palm scan and voice recognition. He stepped into the decontamination chamber, the wind blew and he entered the main lab. He looked around to make sure no one was there. He walked across the room looking at the cameras out of the corner of his eye. He entered the refrigeration room and put on some latex gloves, then opened the cabinet. There were no cameras in the room; they were only pointed at the door. He scanned the cabinet containing multiple petri dishes and vials, each individually labeled. He pulled two small containers, two small syringes and a plastic bag from his pocket and laid them on the counter. He carefully lifted a petri dish from the cabinet labeled BATCH 6AB 20150509. He opened a small vial on the counter, took a syringe and moved some of the live virus from the petri dish to the vial. He capped the syringe and placed it in the plastic bag. Next, he opened a small vial labeled CEREBROSPINAL ST 20150305. He took the second syringe and transferred some fluid from the sample vial into the empty vial, capped the syringe, and placed it in the bag. He placed the original dish and vial in the refrigerator, closed the door and scooped up his samples.

Owen could feel the sweat building on top of his head, so he straightened his hair a little and left the room.

"Find the phone?" the guard asked as Owen stepped through the outer door into the reception area.

"Yep."

"Have a good night, sir," the guard said.

"Thank you. I will," Owen said as he smiled and strolled out of the building.

June 3

"MR. PITKE HERE to see you, ma'am." The young man walked out of the office as Owen Pitke walked in.

"Please, have a seat," the lady said as she rose, shook Owen's hand and pointed to the leather chair in front of her desk.

Alicia Gold was the President and Chief Executive Officer of Antrole Pharmaceuticals. Alicia carried her age well at fifty-five years old. Her black hair was cut short to frame her round, brown face. She had high cheekbones indicative of Indian or Asian descent. Her dark eyes were ominous, but attractive. "What may I do for you?"

Owen cleared his throat and straightened his hair. "I believe you and I have a mutual friend that contacted you yesterday?"

"He did. And, I must say, I am very, very interested." Alicia stood and walked over to close the door. "Shall we move to the sofa and chat?"

Owen stood and followed the woman to the sofa near the window overlooking the Chesapeake Bay. The city lights glistened off the water in the clear night sky. Owen could feel his heart pound as Alicia took her seat. "Yes, I was very interested in what he had to say. Can you elaborate some?"

Owen cleared his throat again. "Yes. Uh, as you know, I am the Director of the Alzheimer's Disease Research Center at Johns Hopkins, so I don't want any of this conversation to ever leave this room."

"Understood."

"Not even to your spouse or friends."

Alicia chuckled and leaned forward to look Owen in the eyes. "I'm not married, and I don't have any friends."

Owen could sense something sinister about her. "Good. I mean ..."

"Please, Owen. Continue." There was no doubt who was in control.

"As Director, I have an … opportunity for you … and me … to … uh, to ……" Owen straightened his hair. "to, uh …"

"To make a lot of money, I think is what you were going to say."

Owen smiled. "Yes, make money." He leaned forward. "A _lot_ of money."

"Then, Mr. Pitke, please tell me of your proposition." Alicia leaned back waiting to hear the story from him.

"We, Johns Hopkins, are experimenting with a … 'cure' for Alzheimer's Disease."

"Yes, that's what I heard from our mutual friend." She seemed almost uninterested.

"I'd like to offer you a sample of our test material in exchange for …"

"Money?"

"Well, not directly. That would seem so crass. I'd rather have some other form of compensation," Owen replied.

"And, what if this 'sample' fails? What assurances will I have?" she asked.

"None. I have nothing else to offer other than it works."

"How do you know?"

"We tested a rat."

Alicia laughed heartily. "A rat? You must be kidding."

Owen straightened up. "A rat. And a maze. Two hundred and seven choices, eighty-four seconds to complete on the second try, no errors."

Alicia's smile quickly vanished. "No errors?"

"None."

"Second try?"

"Yes. And we followed up with two other test subjects; same results."

Alicia took a deep breath and pondered the information. "That is quite impressive."

"Of course, this is only preliminary you understand."

Alicia was offended by his remark. "Of course I understand, Mr. Pitke. We have scientists performing similar tests on our animals as well. I'm not a beginner here, sir."

"I apologize. I didn't mean to offend you."

Alicia composed herself, knowing she may have something significant to build on if she could get it at the right price. "So, what are you looking for, Owen?" she said as she smiled.

Owen straightened his hair. "Well, it seems like a down payment in some form of precious metal and a follow up payment when the product goes to Phase One would be very appropriate."

Alicia thought for a few seconds. "I agree. That does seem appropriate. How much of a down payment are you thinking about?"

Owen had her. He could feel her greed. "Six figures for a down payment, seven when it goes to Phase One."

"Hmmm. You know you'll be able to buy our stock over the next year or so and build your portfolio before we make the announcement. That should give you a nice nest egg in itself."

"It will, but I'd like some assurances up front. What if you are arrested for insider trading or go bankrupt? I need some assurances."

Alicia laughed. "Insider trading. What? You think I do that type of stuff?" She laughed again. "I'm a professional, a very careful professional."

"I believe it," Owen acknowledged. "Regardless, do we have a deal?"

"I need a number for the six and seven figures before I can decide."

"Okay. How about ..." Owen thought for a few seconds. *A yacht, early retirement, a cabin in the Blue Mountains, a world cruise,* "... five and two, respectively."

Alicia thought long and hard. She was a negotiator, and she knew you never take the first offer. "That seems a bit high, Owen, for an unproven product. I can offer five and one as a start."

"Start?"

"Yes," she said. The market for this cure was huge; the potential profit obscene. "When it is proven, we can jump to two."

"Well, that seems pretty reasonable," Owen replied.

Alicia stood and looked out the window. "And what about our mutual friend?"

"Same to him," Owen said.

"O.K. I'll discuss this with my partners and let you know tomorrow if we have a deal. When can you deliver?"

"Tonight."

Alicia turned around to face him. "Tonight?"

"Yep. I have the material safely tucked away for delivery anytime, to you or anyone else."

"Is that a threat, Mr. Pitke?"

"No, not at all. I'm just … available to meet any buyer anytime. I just prefer to deal with your company, if we can."

Alicia settled down. "O.K. How would you like the advance?"

"Yellow is my favorite color," Owen said as he smiled and leaned back in his chair. It felt good to be in charge of such a dominating woman. "Always have liked yellow."

Alicia smiled. "I think we can make this deal tonight with payment tomorrow. Let me make a few phone calls and see what we can do, if you don't mind waiting," she said as she started toward her desk.

"I don't. Not at all." Owen stood and started toward the door. "I think I'll go out and get a bite to eat. Shouldn't take me more than, say, an hour."

"Sounds great. I think I will have everything ready to go by then," she said as she opened a notebook sitting on her desk.

"Terrific," Owen said as he stopped at her desk and to shake her hand. "I must say, it is a pleasure to do business with you, Alicia."

Alicia smiled. "And with you, Owen," she said as they shook hands.

June 12

IMAR WAS WATCHING the young woman prepare the test for Albert, one of the monkeys from Tinjil. Albert was injected with the virus the day after his arrival from Indonesia. That was almost a month ago. He was already showing signs of improved memory retention. Today was a big test for Albert, the analyst and the project as a whole.

A large projection screen hung on supports in front of the monkey, who stood on a small walkway in front of it. He looked at it curiously as the analyst punched some information into the computer. "O.K., Albert." Four pictures appeared on the screen. Two were pictures of a banana. One was of a car, the other a tree. The monkey looked at the pictures as they turned on the screen and disappeared leaving four blank squares. The analyst reached over and touched one square. The square rotated to reveal a picture of the banana. Albert reached over and touched another square, revealing the other picture of a banana. "Good boy, Albert." The monkey clapped as the analyst gave him a grape to eat. The squares rotated back to blank squares. Albert reached over and tapped the two squares to reveal the bananas again. "Good boy, Albert." Another grape. They repeated this several times to make sure Albert understood matching the squares was the way to win a grape.

"Now, let's add some more pictures for you." The analyst entered some data on the keyboard and six more squares appeared. There were three matching pairs. Albert glanced at the squares for a few seconds. The squares rotated to blank squares. Albert touched each square matching them perfectly in sequence, one pair after the other. "Good boy." Three grapes this time.

The pair continued the test uninterrupted.

Imar watched.

—*—

Carl was sitting at the desk when Owen walked in. "Carl, Imar wants us to come down to lab C right now."

"Must be something about the monkey. That's where they are doing the memory tests," Carl said. He grabbed his bag and they headed out the door.

Imar was watching the analyst prepare the final test. The large screen was twelve feet wide and four feet high. It was just the right height for Albert to reach all areas without much difficulty. The screen had two hundred blank squares on it. Albert was pacing back and forth along the walkway in front of the screen, anxious to start the test, hungry for grapes. Imar was standing to the side by the console. Owen and Carl stood behind the rail ready for the show. "Looks like a huge Concentration game," Owen said.

"It is," Imar responded. "We have arranged in random order one hundred pairs of pictures for Albert to match. He has not seen any of these yet and has not attempted a test of this scale. He has successfully completed fifty pairs without error. I thought you might want to see this test live."

"Absolutely," Owen said.

Carl just watched.

The analyst punched in some information and the squares all turned around at one time. Albert scanned the large screen for a few seconds and looked back at Owen and Carl and started clapping. All of the squares turned back to blank sides.

Albert started tapping the monitor with both hands. The squares turned each time the monitor was tapped, revealing matching pairs in sequence. Albert's hands were a blur.

"Unbelievable," Carl muttered.

Albert chattered, jumped, did back flips and clapped as he dashed from one side to the other tapping squares one after another, without error. In less than a minute, the entire board was revealed.

Albert sat and turned to the gallery watching him, clapped and put his hands out for some grapes.

Owen smiled and applauded with the monkey.

—*—

The road to Art Lewell's house was lined with expensive houses ranging from half a million on up. "Quite the hood," Jamille mumbled as he drove to the last house on the left. Pampas Grass in full bloom lined the long driveway that circled around to the front of the brick ranch-style house. Jamille walked up the slate walkway between the two large columns to the front door. "Quite the house for an RV repairman," he mumbled. He rang the doorbell and waited.

Art opened the door wearing shorts and a T-shirt that read *Hard Rock Café—Puerto Vallarta*. "Yeah?"

"I'm Agent Jamille Larson from the FBI." Art's stomach fell. Jamille often saw people respond with surprise that the FBI was calling, but it looked like Art was ready to pass out. He instantly got pale and his hand started to shake slightly. Jamille could tell Art was scared. *What's this guy hiding?*

"Nice to meet you, agent." They shook hands. "The FBI, huh? What brings you here?" Art stood at the door which was slightly open. It was obvious to Jamille Art was blocking the door.

"I wanted to ask you about a storage unit you bought at an auction last year."

Crap! I knew this was going to happen. "I buy a lot of units, agent."

"Please, call me Jamille."

"Sure. I buy a lot of units, Jamille. Which one are you referring to?" *I knew we'd eventually get caught.*

Aha! He's stalling. "One at the Storage Place in Elk Grove. Unit 302B."

He knows! Damn, he knows. I'm going to go to prison. How am I going to survive in there? "Uh, yeah. I remember the unit. Why? Is there something wrong?"

Must be, otherwise you wouldn't be stalling me. "No. I don't think so. I'm just trying to find out what was in the unit."

He knows what was there. He knows about the gold. Crap! What am I going to do? "Wow. Uh, has there been a crime or something?"

"No. I'm just asking a few questions about the unit."

I can't tell him about the gold. "It's hard to remember every unit I buy, you know. I buy a lot of 'em."

Jamille turned and looked at the large front yard. *He's hiding something. If the unit didn't have anything, he would have told me right off.* "Yeah. It must be a good job. Is that why you quit your job at the RV repair shop? Hit a big one?"

I'm going to die in prison. I think I'm going to throw up. "No. Inheritance."

Liar. "Oh. So, about the unit."

"Yeah. Uh, I think that was the one that just had some papers and office crap in it. Nothing real big." *Except the millions in gold that this clown knows about. I need to tell Elias.*

Liar. "So, nothing unusual?"

Gold. Elias will know what to do. "No. Not that I can remember."

"O.K. Well, here's my card. If you remember anything about the unit, can you call and let me know?" *You'll never call.*

"Sure. Is there something in particular you're looking for?" *Like gold.*

I may as well tell him. He isn't going to confess anyway. "Yeah. Maybe something that had some chemicals in it."

Chemicals? Not gold. Are you kidding me? "Chemicals? No, I don't recall anything with chemicals in it." *Except that huge drum.*

"Well, if you do recall, please let me know. It's not a big thing. There is no crime. I just need to know."

Maybe I should tell him if there is no crime. I better talk to Elias, first. "O.K. I'll be sure to do that, agent."

"Thanks."

As Jamille left, Art closed the door and immediately called Elias. "Elias?"

"Art? What the hell are you calling me for? I told you to never call me, you idiot." Elias was furious.

"I just had an FBI agent come to my house asking questions about the locker we bought."

"What? Are you kidding me?"

"No. He was asking about some chemicals, probably the stuff we dumped." Art was pacing by the door.

"You didn't tell him anything, did you?"

"No. He said there was no crime; that he just wanted to know."

"I don't believe him." Elias knew he had to distance himself from Art right away. "Are you sure you didn't tell him anything else? Anything about me?"

"No. I didn't say anything."

Elias knew he had to make sure Art was the only person tied to that locker and that it stayed that way. "Where are you?"

"At home."

"O.K. Let me think." The phone was silent for a few seconds. "Let's meet at the fishing hole and figure out what to do. Be there in an hour."

"O.K." Art placed his phone back in his pocket and continued to pace the room.

Jamille sat in his car down the street and watched the house in his rear view mirror. He saw Art walk outside, get into his pickup and back out of the driveway. Jamille scrunched down as Art drove by, never noticing him. Jamille noted the license plate, color and type of truck. He waited until Art was almost out of sight, started his car and followed him.

Jamille was excellent at tailing someone. He knew the area and the roads. He also knew the key to being successful was staying in the person's blind spot and out of sight as much as possible.

Art stopped at a gas station, then at a grocery store. Jamille thought it might be a waste of time, but he was going to stick with it for a little while, just in case.

Art left the store and headed south toward the Stockton Delta on the country roads. Jamille followed at a distance and watched as Art drove across the hills and fields. It was easy to follow him since he could see him at a distance. Many of the roads were empty; just Art and Jamille, who followed at a great distance and occasionally turned off on long straight-aways, but kept an eye on Art.

Jamille stopped on an overpass and watched Art as he drove to a levee on the Mokolume River near the town of Hood. Suddenly, Art turned off the levee and disappeared. Jamille accelerated and was in the area where Art turned within a minute, but couldn't see his truck anywhere. The only place Art could have gone was down the levee toward the river. Jamille parked the car and walked over the levee for a peek. He stood behind some trees and scanned the levee banks. He saw two vehicles; Art's truck and another truck. He didn't recognize the man with Art; he was too far away. Jamille watched as Elias and Art talked.

The two men were standing near the water discussing something. It appeared they started off just talking and soon were raising their voices and becoming more animated. Jamille could hear them but couldn't understand anything that was said. He's seen people argue before and this looked like it was quickly becoming a heated argument. Art turned away as the other man was shaking his finger at Art. Art was waving his arms like he was lost or giving up or something. Jamille watched as Elias grabbed something off the ground, maybe a rock, and suddenly ran up behind Art and smashed him over the head with it. Art instantly crumpled to the ground and the man stood over him, still yelling and pointing.

"What the ...?" Jamille pulled out his pistol and started walking toward the men. Jamille figured they were three hundred yards away and it would take him a couple of minutes to cover the ground. He pulled out his cell phone and called 911.

"Nine one one. What is your emergency?"

"This is Agent Jamille Larson from the FBI. I am witnessing an assault with a deadly weapon and injury at Hood River Rd mile marker fourteen south of Elk Grove, west levee bank of the Mokolume River. I need back up and medical assistance immediately. Badge number 13786."

"Please stay on the line ..." Jamille hung up as he continued to move toward the men. He stopped behind some bushes to see what was going on. Art lay motionless on the ground. Elias grabbed Art by the legs and drug him to the river's edge. With a heave and kick of the legs, Elias rolled Art into the river. Jamille watched as the man took the large rock and placed it on the edge of the bank. Then, he slid his foot next to the bank, apparently to make it look like Art slipped. He walked over to his truck and pulled out a fishing pole, tackle box, and a bag. He put the bag on the ground and opened the tackle box and set it down. Then he cast the line from the pole into the river, reeled in a few times, and tossed the pole to the edge of the river where Art was floating face down.

Jamille decided he needed to get closer after he watched the scene set up. He was trying to give his backup a little time to arrive, but he couldn't wait any longer. He knew the guy was going to leave. He was almost there; just a hundred yards away.

Elias grabbed a branch and swept the drag marks from the dirt. He tossed the branch into the bushes, looked around and, satisfied with his scene, started toward the pickup.

Jamille was seventy-five yards away and standing behind a large pile of blackberry bushes. He'd have to go fifty yards in either direction to get around them. He could see the road down in front of him to the right. Elias would have to drive right by him to leave the crime scene. Jamille heard Elias start his truck. Jamille decided he had to move quickly, so he bolted down the bank behind the blackberry bushes and stopped on the side of the road, waiting for the truck to pass by. He pulled out his badge and his gun and stood at the end of the berry bushes on the side of the road. He didn't hear any sirens. He knew he was alone. He knew Art was dead. He knew he had to stop this guy if he tried to leave.

The truck rumbled down the road toward him. Jamille's heart started to pound as the truck got closer. Jamille watched as the truck bounced through a couple of potholes and rounded the corner. Jamille took a deep breath and stepped out from the berry bushes about sixty yards away from the truck, held up his badge and pointed the pistol at the driver. "FBI. Stop, or I'll shoot," he shouted. He was looking down the barrel of his pistol into Elias's eyes.

Elias was startled and didn't know what to do. He reacted and pushed the gas pedal to the floor. The truck accelerated and lunged toward Jamille. It hit a huge pothole and jumped several feet into the air, breaking the front axle when it crashed back to earth. The truck veered to the left as Jamille dove to the right. Jamille watched as the truck plunged down the levee and into the river head first and started sinking. Elias crawled out of the window and started yelling, "Help. I can't swim." He was yelling and gasping for air, slipping under water between words, coughing and choking.

Jamille watched as the truck quickly filled with water. He thought he might let the guy drown since he just tried to run him over, but he realized he couldn't. He holstered his gun and ran downstream a few yards as the truck slipped under water and Elias floundered. Jamille waded into the edge of the meandering river about waist deep as Elias was struggling by. Jamille held onto a willow branch with one hand and leaned out into the river. He reached out to grab Elias's arm and saw Art float by about thirty feet away face down. Elias grabbed Jamille's arm as he started to float by. He clung to his arm, coughing and gasping for air. Jamille pulled him toward the shore and glanced downstream, but couldn't see Art any longer. Jamille hoisted Elias onto the bank and pushed him facedown into the dirt. Elias was coughing and choking. "You'll live," Jamille said as he handcuffed Elias's hands behind his back. "You're under arrest," he said as he put his knee on Elias's back and checked his pockets for weapons. None. He sat him up and leaned him forward so he could continue to cough up some water. "You can stay here or run, I don't care. If I have to chase you again, I'll be pissed and I won't be as forgiving." Jamille looked at the river. "I wouldn't jump in with those bracelets on." Jamille could hear the sirens approaching in the distance. "Company."

Jamille rushed downstream a short way, but couldn't see Art. He suspected he went under. He walked to the top of the levy and waved to the advancing police car. He walked over to Elias as the police car bounced down the levee road. "What's your name?" he asked.

Elias spit out some dirt and water and shook his head. "Does it matter?"

"No, not really. We'll find out anyway once we charge you with murder."

"Murder? What are you talking about?"

Jamille stood over Elias. "I saw you hit Art with a rock and toss him into the river. He just floated by us while you were in the water squealing like a little girl. We'll find his body, and the rock, and your fingerprints or DNA on it"

Elias was shocked that an FBI agent would know his friend. "How do you know Art?" still coughing.

"I don't. I asked him a few questions about the storage locker he bought last year."

"Storage locker?" Elias knew he was trying to find out about the gold.

"Yeah. And next thing you know, he ends up here and you kill him. Trying to hide what you did with the chemicals?"

Elias couldn't believe what he was hearing. "Chemicals? You want to know what happened to the chemicals?" Elias started laughing. "Chemicals." He fell over onto his back as the police officer walked up.

"What's so funny?" the officer asked.

Elias struggled to catch his breath between his laughter. "I thought you were nosing around about the gold."

Jamille squatted down next to Elias. "Gold? What gold?" Jamille asked as he helped Elias to a sitting position.

Elias looked at Jamille. "The gold we found, dumb ass. Who cares about the fertilizer. We dumped that in the sewer."

Jamille stood to his feet. "Huh. Gold." He started to walk away and turned back to the officer. "Better read him his rights. And get a dive team here to search for the victim. I saw him over there a few minutes ago headed downstream, facedown." Jamille started to walk away as the officer pulled Elias to his knees. Jamille stopped, turned back, squatted next to Elias and whispered in his ear. "You shouldn't call people bad names." Jamille flicked Elias's ear causing him to squeal.

Jamille smiled and walked off listening to the officer reading Elias his rights.

Phase One

September

CARL'S WORST FEARS had been realized. Jamille's discovery that the chemical responsible for the 19Q disaster several years prior was found and dumped into the sewage system by two yahoos was extremely troubling. It was dificult enough to think that this could create another mutated organism that could be released into the environment and could kill people, but to find that it created a mutated virus that cured people was bizarre. He couldn't accept it. How could something so bad end up being so good?

But everything pointed to just that. Two of the three patients that started the research, Mac and Quinton, were doing just fine, mentally. The autopsy revealed Bonnie died from pneumonia attributed to the leukemia; not the virus, which was a relief to Owen. Mac was singing at various functions and giving speeches about honor and his service during the big war. And Quinton, well, he was already a world class poker player and came in third in one of the biggest poker tournaments in the world, earning a cool six million dollars. The only drawback for Quinton was he divorced his wife to seek his fame and fortune. She was a hindrance to him anyway, so he thought.

Neither suffered any adverse side effects from the virus. Quinton's migraines were cured when the doctors discovered he developed an extreme allergy to aspartame; an artificial sweetener that is metabolized into the body as phenylalanine. The virus could have been the cause for the new migraines since phenylalanine is one of twenty common amino acids used to biochemically form proteins, but it didn't matter. They discovered the allergy and changed his diet and the migraines stopped.

And Quinton was a millionaire.

Because he worked at the sewer plant and had a cold.

The day after Jamille contacted Carl about the sewage dump, Alice and Carl did some online research on the plant where Quinton Lemolo worked and determined it was within a few miles of the dump site. They contacted the plant manager and explained they were doing some research on viruses and bacteria found in sewage treatment facilities. The manager confirmed the plant was fed from the industrial park through pump station six on the north side of Elk Grove. It would take about twenty-four hours for the residue to reach the plant for processing. That fit their estimated time frames of when Quinton was seriously ill at work.

They discovered that the plant was designed as a reconstructed wetlands for discharge into the adjacent Applegate Wetlands. They plotted the location of the school and the Golden Pond Memory Care Facility, both of which also happened to be adjacent to the wetlands.

Carl flew to Sacramento with an analyst and gathered samples of water from the sewage plant and the wetlands to analyze and hopefully confirm their suspicions; that the virus was alive and well in the water. They met with the Elk Grove Wastewater Treatment facility manager, James Bulger. He was a stout man in his early sixties. He knew the plant better than anyone, having been there nearly thirty years and personally supervised all of the expansion and development into a constructed wetlands discharge. He knew everything about the plant, and very little about why Quinton suddenly quit. James thought it was sad that he divorced his wife after having their first child. Sad.

Carl then toured the Golden Pond Memory care facility with the administrator, Melanie Grimes. She was an impressive woman in her late fifties and well educated. Her compassion toward the elderly with dementia was contagious. She recalled Stanley Thorne and was proud to share the success of his recovery. In fact, she had two other residents that were showing signs of improved memories as well. Carl left with their names and a sense that this was ground zero for the virus.

Next was the school where little Bonnie Howe attended. The school was just breaking for summer vacation so their timing was perfect. Teachers were laid back and very willing to discuss the curriculum and students. The principal knew of Bonnie's participation in the test group and was eager to share about her successes. She relayed it had to be partially the result of the school's excellent focus on the environment for the third and fourth graders, which included tours to Applegate Wetlands and the Elk Grove Wastewater Treatment facility. She was saddened by Bonnie's recent death.

It all fit. Three locations all tied to the wetlands.

Carl just needed to get the samples back to the lab for confirmation. When he arrived back in Maryland, his team was ready to take the water samples and start the research. He explained some of the difficulties he had trying to get the samples on board the flight to Maryland, but after a few phone calls and several faxes, he was able to bring the quart of water from the wetlands to the lab for a thorough analysis.

Carl and Imar had just assigned tasks to their team when Owen walked in, obviously excited. "Carl, glad to see you're back. I need to talk with you and Imar in the conference room." He didn't wait for a response as he walked into the conference room and sat at the head of the table. "I met with the board while you were gone. We agreed that this needs to get on the fast track for FDA approval. I've contacted Don Sinclair at the FDA to work with us to get the CDER[19] to agree to get Phase One testing going."

"We met Don when we worked on the 19Q project. Seems like a good guy," Carl replied.

"He is," Owen confirmed. "After I showed him the results of some of the tests with the animals, particularly Albert, he was gung ho with this."

"Gung ho?" Imar asked.

"Yeah. It means in agreement, like moving forward. A term the Marines would use." Carl explained.

"Oh. Complicated," Imar said.

"Anyway, he wants to work with us to get this on the fast track. It's not going to take us ten years to get this to market. He's saying two."

---

19 CDER—Center for Drug Evaluation and Research—FDA department responsible to oversee and approve the development of new drugs.

"Two?" Carl asked. "That's unheard of."

"I know. He sees this as being the cure of the century and wants to be in on the ground level. He said he can make it happen, and if anyone can at the FDA, it's Don."

"What does he need from us?" Imar asked.

"I need to have you work with our product development team to put a Phase One test group together. They will have all of the details of the sample size, test subjects, waivers, all that stuff," Owen said as he made sure his hair was straight. "They've worked with the FDA many times before and know what to do to get this to CDER. They will work with Don to arrange meetings with the FDA's Central Nervous System Drugs Advisory Committee to get this going." Owen stopped playing with his hair and turned serious. "Carl, I want this to go to market as soon as we can. It's going to be a huge boost to our program and funding."

"Owen, I've discovered some things about this virus that are … troubling," Carl said.

Owen leaned forward. "What do you mean 'troubling'?"

"I've confirmed the virus was mutated by the chemical Everett Maskill created in the early 2000's. The stuff that resulted in the 19Q disaster. It's a mutated virus."

"So what? That doesn't matter. It works. It's curing people, not killing them. Who cares why it works? It just does."

"Believe me, this can't be a good thing. That stuff is deadly. I feel it."

Owen slammed his hand on the table. It startled both men. "Listen. I don't care what you 'feel,' Carl," he said cynically. "We have the greatest cure of the decade here, maybe the century. It works. Get over your fears and get this to market or I'll find someone who will." Owen stood to his feet to look down on Carl and Imar. "Get with the product development team and the FDA and get this to Phase One." He stormed out of the room.

Imar looked at Carl. "He has a point, Carl. It works."

Carl was silent.

—*—

Bringing a new medicine to market is no easy or simple task. It normally takes years for a product to be tested and approved by the FDA. Just the first phase of animal testing can take several years. For Owen and Johns Hopkins to get the FDA to agree to human testing in a Phase One clinical trial after only six months of animal testing was a phenomenal accomplishment. It was made possible by the long hours worked by every member of the team to document the animal tests for the FDA and submit the results with the New Drug Application which was approved in just three weeks. Don Sinclair pulled through as expected. The application provided the authority for Carl's team to establish a Phase One test of the new drug; Cerebtol.

Owen, Don Sinclair and Imar agreed the test group should be forty to fifty people from diverse geographic areas. They would be from different sexes, ethnic groups and medical backgrounds. The only common factor amongst the test group would be the diagnosis of advanced Alzheimer's disease. The team would need to inform each patient of the risks associated with the test and obtain their consent. Don thought this would be problematic since each participant was diagnosed with advanced Alzheimer's and did not have the capacity to understand the risks associated with the tests, much less provide consent for the treatment. Carl suggested using only patients who had court appointed guardians. The guardians would be able to provide consent on behalf of the patient. They did, and by the end of the month they had a test group of fifty people in seven states willing to participate in the Phase One clinical trial.

Year Three—2016

Phase Two

April 18

IT WAS A beautiful spring day in Maryland. The rains stopped just long enough to let the sun warm the ground causing small whiffs of steam to rise into the sunlight. The birds were singing and the flowers were showing their blooms. Carl Kruger could smell the honeysuckle plants as he walked up to the front door of the research center at Johns Hopkins. He took the same route every morning and had done so for years. Today, however, he was apprehensive about approaching the building. He didn't really want to go into the meeting with Don Sinclair, Owen Pitke and Imar Spaan to discuss the results of the Phase One clinical test of Cerebtol. He would rather be fishing with his seven year-old son, Joey, or spending the afternoon at the classic car show at Riverside with Alice and Joey. Anything but this meeting seemed better at the moment. Maybe it was because of the change in weather. Carl couldn't put his finger on it, but he just didn't want to be there to discuss the results of the testing. Not today.

The Phase One clinical trial was a huge success. Fifty patients from various states were selected to participate in the Cerebtol test. Of the fifty patients, all showed significant signs of neurogenesis. Their memory retention improved significantly on an average of twenty percent. The scans showed regenerative areas of the brain in all cases. No one suffered any side effects. No one had difficulties with receiving the virus. All of the patients were facing discharge at their respective facilities. One patient died from a stroke. Two developed pneumonia and died. Two had heart attacks and died. None of the deaths were associated with the medicine; just aging issues. Overall, it was a tremendous success.

Owen was standing at the head of the table adjusting his hair when Carl walked in. "Good morning, folks," Carl said.

Owen smiled. "Yes, it is, Carl. Yes, it is." Owen started right in as Carl took his seat. There was no time for chit-chat. "We're ready to go to Phase Two, Carl."

Carl was surprised. "Already? Phase Two?"

"Yes," Don replied. "Owen showed me the results of Phase One and it is quite impressive. I know I will be able to get the committee to approve a Phase Two trial right away."

"But, Don. Don't you think we should do some more tests to make sure we haven't missed anything?" Carl asked.

"We did the tests, Carl," Owen retorted. "What do you think we've been doing here for seven months?" Owen was obviously perturbed by Carl's question.

Imar came to Carl's defense. "I think what Doctor Kruger is trying to say is …"

"I don't care what he is trying to say, Imar. What I care about is getting this product moving, and if Don agrees the tests look good enough to propose a Phase Two test, then we need to do it."

"Owen, you don't need to get all upset about this," Carl said.

"I'm not upset!"

"Sounds like it to me," Imar said.

"Listen. This is serious. I don't need to have you try and slow this down with your fears. That's all I've been hearing for months. 'It's mutated.' 'We should do more this or that.' Carl. People are suffering out there and we have the cure. It works. It works flawlessly. Don agrees. End of story," he said as he tossed a thick report on the table labeled "Cerebtol: Phase One Test."

"Carl. I understand your concerns," Don said calmly. Owen was turned away and straightening his hair as Don spoke. "I went through 19Q with you. I remember it very well and I certainly don't want to experience anything like that again. Suggesting the bombing of Quincy was absolutely agonizing. That's why we are being very careful about this while working expeditiously.

"The purpose of Phase One was to validate the drug's safety and establish a profile. We did that with no events of any type." Don stood and walked to the end of the table near Carl and Imar. "None. If we move to Phase Two, we will be able to continue the validation while we examine the efficacy of the drug. We will have plenty of time and research before this goes live." Don leaned forward and placed his hand on Carl's shoulder. "No one wants anyone hurt from this. In fact, we just want to see people healed," he said with a broad smile.

"I agree," Imar said.

Owen turned back to the group, pleased that Don had persuaded them of the benefit to continue to Phase Two. "So, Don will get the report to the committee for review and propose commencement of Phase Two testing. Agreed?" Heads nod. "Great." Owen grabbed his papers and report and stuffed them into his briefcase.

Imar hesitantly raised his hand. "Uh, Owen."

Owen stopped, surprised since Imar seldom interrupted anything. "Yes?"

"I think my work here is done."

Carl was also surprised. "Done?"

"Yes," Imar said. "We have succeeded in a Phase One test, and are moving into Phase Two. It seems to me that any adjustments you make at this point can be done without my presence."

"You have been here quite a long time, Imar," Owen said.

"I appreciate the vacations, Mr. Director," Imar said. "However, I believe it is time for me to return home, to my country."

"Nothing wrong, I hope."

"No. Nothing like that. It's just time for me to go." Imar looked at Carl. "I miss home."

Carl put his hand on Imar's shoulder. "I understand. I think our work here is done. Well, your work is." He paused. "I still have a lot to do, but it's just oversight at this point. What if we need your advice on something, Imar?" Carl asked.

"Then you will have it. In today's world of electronic information and communication, we can still be very connected, Carl." Imar smiled. "Very."

"I agree," Owen said. He walked over and shook Imar's hand. "Thank you for being such an integral part of this team, for your sacrifice and commitment. We, Johns Hopkins and, to be frank, the world, are grateful for your contribution."

Imar shook each person's hand as Owen continued to put papers in his briefcase.

"When will you leave, Imar?" Carl asked.

"Oh. Maybe Friday. That will give me some time to wrap up with my team."

"Then, we will have a little celebration party before you go," Carl said.

"Yes, we should," Owen agreed.

"Will I be invited?" Don asked.

"Of course," Owen said. "Now, if you gentlemen will excuse me, I have another meeting to go to." Owen grabbed his briefcase and headed out the door.

The meeting was over. Phase Two was ready to commence. Five hundred people would soon receive their injection of Cerebtol.

The Performance

December 24

STANLEY MACKENNA THORNE was standing offstage peeking through the curtains at the crowded auditorium of the Elk Grove Faith Center. There were more than six hundred people anxiously awaiting Mac's annual performance. To hear him sing was a treat. To know he was ninety-three years old was inspiring. To know he was a Congressional Medal of Honor recipient was humbling.

Mac watched as the animals on stage seemed relaxed. He wondered if they were given sedatives to calm them down with all of the people and noises nearby. Usually, animals get skittish when loud noises occur, like a crowd applauding the piano recital of the Twelve Days of Christmas by seven year-old Emily Jansen. The last round through the Twelve Days her hands were a blur. All eyes were on Emily; even the animals, or so it seemed.

Mac was due to go on right after Emily and just before the choir. The Everland Stringed Orchestra, also regulars at the performance, would accompany Mac playing 'Oh Holy Night,' just as they had years prior. This year there were no introductions. The pastor wanted the evening to flow without interruption so the audience could feel the emotions of each per-formance without a break. As Emily took her slight bow and started to walk off stage, Mac started onstage to the roar of the audience applause. The contrast of the elderly black man and the little white girl was stark when

Mac stopped mid-stage and knelt down to hug Emily. The little girl reached up and gave Mac a big hug and kissed him on the cheek, then continued offstage.

The audience rose to their feet, applauding heartily.

When the audience quieted down and reseated, the lights dimmed and the orchestra struck the first note. As usual, the music was flawless. Sounds floated through the air like a sweet fragrance on a spring day. Mac took a deep breath and started to sing the song effortlessly as he always did.

"Oh Holy night ... "

His deep voice resonated through the auditorium and melded perfectly with the orchestra. The performance was inspiring.

"... The stars are brightly shining ... "

Everyone was fully engaged with the song. Mac's voice was mesmerizing. To watch him sing to the heavens with arms expressing his praise, dressed in his tuxedo, with such a magnificent voice was a divine moment for many.

"... It is the night of our dear Savior's birth ... "

The orchestra continued to play the song to the next line, but Mac didn't say anything. He stopped singing and the orchestra continued. Mac's eyes opened and he saw the audience looking at him. He started to get a sense of fear. He could feel he was breaking out in a sweat and his hand started to tremble.

He forgot the next line.

He looked scared as he hummed the melody and the orchestra continued to play, until he heard someone backstage say the words and he remembered.

"... A thrill of hope the weary world rejoices ... "

Mac continued to sing the remainder of the song without error. When he finished, the sanctuary burst into applause and cheers as everyone stood to their feet. Mac hit the high note perfectly and the audience relished in it and gave due credit. Some people were crying, many applauding heartily. Mac scanned the audience and had a strained smile on his face and a tear rolled down his cheek. A tear not from joy of his performance or humility from the audience appreciation, but a tear of fear.

He forgot a line to a song he had sung a thousand times before.

Mac lowered his head and slowly walked offstage.

—*—

The street in front of the apartment was dimly lit as the fog stole the light from the night. It was cold and damp, and Mac could feel the chill to his bones. He felt terrible. He was wrapped tightly in his wool coat and could see his breath as he climbed the short set of stairs to the front of the small apartment building. The lights flashed along the fence, changing colors and rhythms in unison with Mac's stride. He walked up to the bright red door that his neighbor, Janice, wrapped to make it look like a Christmas present. He didn't notice that there were decorations everywhere; the bush decorated like a Christmas tree, the wreath on the door, the plastic snowman near the mailbox. Mac was thinking about his performance at the church. He was embarrassed that he forgot a line in the song. More so, he was scared.

Mac had been scared before. When he charged the enemy in Normandy, he was scared. Men were blown to pieces around him. Yet, he endured. When his wife died, he was scared; scared to be alone, scared of what the future held. Yet, he endured. His faith always pulled him through. But tonight, he was struggling, mightily.

He didn't want to go back to Alzheimer's hell.

The thought terrified him. It was out of his control. In the war, he could make choices. In his life, he could make choices. Yes, some issues were beyond his control, but the ability to manage through them was his to own. With Alzheimer's he had no control. Nothing he could do would prepare him for the inevitable.

Mac walked into his apartment and slowly closed the door. "Tony?" No answer. The place was empty except for the cat. *Tony must be at his girlfriend's parents' house,* he thought. He flipped on a few lights, turned up the furnace and sat in his chair. His cat, LuLu, jumped into his lap and started to purr. Mac stared straight ahead as he slowly stroked the back of the Tabby cat as it nestled against his chest and tried to climb into his coat.

It had to be a mistake, a one-time thing, he thought. Mac placed LuLu on the floor as he stood and walked briskly to the kitchen. He rummaged through a junk drawer and grabbed a cassette recorder. He grabbed a recipe book from the counter and sat at the table, knocking over a glass and breaking it. Mac paid no attention to the shattered glass as he opened the book and randomly turned to a recipe. He read the recipe out loud and then closed the book. Mac turned on the cassette recorder and recited the recipe.

When he finished, he opened the book, rewound the recorder and played it back. He followed along with his finger as the recorder played back the recipe Mac quoted. His hands were sweating. LuLu tried to jump into his lap, but he shoed her away.

"… one-half teaspoon of cinnamon, one quarter teaspoon of cumin, a dash of salt. Mix ingredients in a large bowl. Fill the pie crust and bake at four hundred degrees for …" Mac stopped the recorder. He rewound it. "… crust and bake at four hundred degrees …"

Click! He stopped it again, and rewound it.

He continued this several times.

Click!

"… four hundred degrees …"

The fifth time, he closed the book, lowered his head, and sobbed.

LuLu jumped into his lap to comfort him.

Year Four—2017

Breaking News

February 6

THE WIND WAS blowing steadily outside with strong gusts. Carl was standing by the window watching the trees bend in the wind, wondering if any would blow over. He could barely see the road because of the blowing snow. "What a miserable day," he mumbled. His throat hurt, he had a slight fever and he just did not feel like being at work. It was mid-morning and he had already spent several hours reviewing some reports on the progress of the Phase Two test at home. His head was starting to hurt, so the view out the window of the winter storm was, oddly enough, a welcome sight.

The initial Phase Two tests were going remarkably well. There were no adverse side effects with any of the patients. None. All four hundred and eighty-three patients were showing marked signs of memory improvements. Most were elderly. Thirty-two were middle aged with Early Onset Alzheimer's, the most aggressive type. It seemed their progress was a little slower than the other patients who were elderly with advanced Alzheimer's, but their progress was remarkable none the less. All of the patients were communicative. All were able to manage their daily activities with minimal assistance. Seventy-six patients were discharged from facilities. Nineteen were taken off guardianship. And they were only on the medication ten months.

The house was quiet as Carl strolled back to the sofa and sat next to the pile of reports. Joey was off to school and Alice was at her sister's house doing who knows what. Carl glanced at the remaining reports but couldn't get the juices flowing to start the review process again. "Stupid throat," he said to no one. Carl decided to flip on the television and see what was happening in the business world.

A middle-aged woman was sitting at a table next to an older man who had a full head of hair, nice tan and was a little overweight. She was reading something that Carl couldn't hear. He fiddled with the TV and realized the surround sound was muted. "Stupid controller. Everything's just stupid today." He kicked his shoes out of the way. "Stupid shoes."

The volume went up on the TV. The woman was speaking. *"... much more so than anyone would imagine. The volume is topping four million shares in the first twenty minutes of trade. This is the largest single day increase percentage-wise on record and it's just started."* Carl watched intently, drawn by the excitement of the news. *"The stock is up to fifteen dollars and eighty-five cents per share since the market opened twenty minutes ago; that's an increase of six thousand three hundred and forty percent!"*

Carl adjusted his glasses and leaned forward. "What the ...?" he mumbled.

The man on TV took over. *"That's right, Maria. Antrole Pharmaceuticals is breaking records today and looks to be the hottest stock of the last fifty years. The pharmaceutical company announced at the market open that they were developing a cure for Alzheimer's and that Phase One of their clinical test was approved."*

"What?" Carl blurted.

The man continued. *"This unknown penny stock of yesterday is about to become the golden stock of today. When you look at the number of people with Alzheimer's, the fear associated with that terrible disease, and what their product is capable of doing, there is no wonder that this is the hottest stock on the market."*

Carl searched around for his cell phone as the commentators continued. *"They apparently are able to use a virus to stop the spread of the ..."*

Carl froze and stared at the TV. "Virus?" he shouted. "Virus?"

"... even is capable of reversing some of the damage caused by the disease."

Carl was near frantic. He was tossing pillows looking for his phone in the cushions as the commentators continued their coverage of Antrole Pharmaceuticals stock. He found his phone in the folds of the sofa, grabbed the controller and muted the TV as he dialed frantically. The phone rang several times until Owen's voice came on asking the caller to leave a message.

"Owen, this is Carl. There's a report on TV about a company testing a virus for a cure for Alzheimer's. Did you know about this? What is going on here? I need to talk to you right away. Call me when you get this. On my cell." Carl tapped the screen to disconnect. He looked at the reports and back to the TV, watching the ticker on the stock scroll across the bottom of the screen with charts and graphs of the price climb. "Damn." He grabbed his coat, keys, and headed out the door to see Owen.

—*—

As Carl approached Owen's office, he could hear him on the phone. "... on the ground floor. Can you believe it?" Owen chuckled as Carl walked in, unaware of his entrance. Owen was standing at the window looking out at the approaching storm talking on his cell phone. The wind was blowing so hard, the windows rattled every few seconds. "Amazing. I had no idea. It's better than I ever imagined." Owen straightened his hair and turned around to see Carl. He had a huge smile across his face, and suddenly seemed embarrassed. "Uh ... I gotta go. Yeah. I'll talk to you later. Ha, ha. Sure." He tapped his screen and put it in his pocket. "Carl," he said as he smiled.

Carl was obviously upset. "Did you know about this company testing the virus?"

"What?"

"Antrole Pharmaceuticals. A company testing the virus for a cure for Alzheimer's. Did you know about this?"

"No. Of course not. What are you talking about?" Owen asked as he moved to his desk and straightened his hair.

"It was on the business news. Some company is testing a virus for a cure for Alzheimer's and their stock has gone up thousands of percent in the first half hour of trading."

"Huh." Owen didn't seem interested, or even concerned, which surprised Carl. "Do we know if it's the same virus?"

"No! Of course not. How would we know? I just heard about this less than an hour ago."

"Carl. There's no need to get upset," Owen said, trying to calm Carl down. "It doesn't surprise me someone else found out about it."

"What?"

"We know there were other cases starting to surface, so it was just a matter of time before someone found out and discovered the virus."

Carl seemed to settle down. It made sense. Sure. Someone else was bound to discover this thing and move on it, just like they did. "I ... I never thought about that." Carl sat in the chair at the table. "I thought ... I thought ..."

"What? You thought I knew about this? Carl," Owen said reassuringly. "We're in this together. If I find out something about this virus, I'll be sure to tell you and Imar right away. We're a team." Carl smiled slightly as Owen slapped him on the shoulder. "So, tell me more."

"All I know is the company; Antrole Pharmaceuticals."

"O.K. I'll look them up and see if they will share what's going on with their study. If it is the same virus, this could benefit us."

"How so?"

"I think it could be wise to partner with them to bring the product to market. Their participation in development provides additional resources and an avenue to distribute the product worldwide when it goes to market. It could benefit both of us."

"Wouldn't that reduce the revenues the university could obtain from the sale of this?"

"Oh, I'm sure it would, but how much? I mean, this is a huge issue, and there is a tremendous demand for a cure. I think as long as we get credit for first discovering it and we are the first to bring it to market, then having another company tag along could be very good for us. It would show we are willing to share our knowledge for the betterment of man," Owen said as he spread his arms wide.

"Sounds a bit premature ... to enter into a partnership."

"Look. We'll still get credit for the creation of the product, which will be a huge boost to our image and funding. I expect our endowment program will double. People want to be associated with success and finding a cure for Alzheimer's is tops. Our alumni will give millions to have their legacies associated with Johns Hopkins University; the creator of the cure for Alzheimer's. That alone is success in my eyes." Owen smiled as he thought about the success, the press, the opportunity to go down in history as the institution that found 'The Cure.'

Carl watched. He was preoccupied with his cold, the news, the storm. His mind was wandering as Owen droned on.

"So, what I need to do is contact these people and see how far along they are in Phase One. Then, I'll discuss this with the board and see if they agree to bring them into our testing while we ask the FDA to make the announcement of our discovery. We just need to make sure we are a step ahead of them all the way." Owen noticed Carl wasn't feeling well. "You don't look too good."

"No. I wasn't planning on coming in, and then this news …" Carl said as he coughed.

"You better get back home and get some rest. We can pick this up tomorrow. There's a lot of bad viruses out there and I don't want you to pick one up," Owen said as he smiled, humored by his little joke.

"Yeah. O.K. I'll see you tomorrow," Carl said as he walked out of the office.

Owen turned and looked out the window again at the storm. "Sweet," he whispered.

At the Park

April 23

IT WAS A beautiful spring morning at Elk Grove Park. The giant, valley oak trees, full of leaves, branched across the roads with a canopy of green. The park was created back in 1903 when the property was first purchased to harvest the trees for lumber and fuel. A young woman named Jennie McConnell wanted to preserve the trees, so she decided to sell shares to the citizens of the community for $5 each in order to purchase the property back and turn it into a park. Imar and Carl appreciated the efforts of this woman they never met. "These trees are amazing, aren't they?" Imar whispered to Carl.

"Yes, they are, Imar. Oak trees. Pretty majestic, I must say," Carl whispered back.

The two men were standing near the rear of a group of people watching the soldiers, dressed in their formal attire, carry an urn and a folded flag to the front of the assembly. The urn was placed on a pedestal and the two soldiers stepped to the side and stood at attention. Many of the people in the audience were elderly. News cameras rolled at the back and sides of the event capturing the moment. The Vice President and Secretary of Defense sat in the front row. Other dignitaries sat in the front two rows, including Major General Ritchey.

It wasn't every day our country puts a Congressional Medal of Honor recipient to rest.

Stanley MacKenna Thorne always wanted his memorial to be in Elk Grove Park. He loved the park. He spent many days there with his family after leaving the military and settling in this once-small town. He used to take his kids to the park pool, have picnics there regularly, watch the horse shows and parades of the Elk Grove Western Festival, the Fourth of July fireworks and so much more. His fondest memories were centered around the park and that was where he wanted his friends and family to say goodbye.

Several dignitaries stood one at a time and gave a word of thanks for Mac's service, his bravery, his humor. Carl was surprised to hear that Mac killed so many people in the war. He had a difficult time picturing the old man as a young, strong, deadly soldier who charged machine gun nests risking his life to save his fellow soldiers under fire. What the war must have been like, he wondered. How terrible.

Several great-grand children stood and talked about their 'Grandpa Great' and the influence he had on their lives. Mac's grandson, Tony, told about Mac's acceptance of Christ, the last days when he and Mac lived together and how much he loved being able to live on his own again and how Mac was a rock for him during his trials in life. He told of Mac's love to fish, and when he bought Mac a lesson to a fly tying class last Christmas, and when Mac was excited about it and kept saying he couldn't wait to go to the 'tie flying class' with Tony. He just couldn't say it right. How funny, he said.

Carl and Imar looked at each other.

A uniformed man stood to the side of the audience. "Atten hut!" the Master Sergeant barked. "Ready ..." Seven armed marines loaded their rifles. "... aim ..." They brought the weapons up to aim. "... Fire! ...... Fire! ...... Fire! ..." Carl and Imar flinched at each shot. They had never seen a twenty-one gun salute live and were surprised at how loud the rifles were.

A corporal played taps. It was flawless.

The two soldiers stepped in front of the pedestal and unfolded the flag. They held it at full length in proper format in front of the urn display-ing its blue field of stars and red and white stripes, then snapped it to a flat presentation and began folding it in front of the audience in military fashion. Once the folding of the flag was completed, a soldier knelt in front of Tony and handed him the flag. "On behalf of a grateful nation and a proud Marine Corps, I present this flag to you in recognition of your grandfather's years of honorable and faithful service to his country." The soldier then stood, stepped back one step, and saluted Tony. He turned and joined the other soldiers at the urn.

One soldier lifted the urn and carried it to a black, horse-drawn Hearst carriage, followed by the other soldiers. Imar and Carl watched as the horse trotted away with several soldiers walking slowly behind the carriage. They would follow Mac's remains down the main boulevard of Elk Grove. Later, Mac would be loaded into a car and taken to the airport for his last trip; Arlington National Cemetery where many of the same people would attend a second service for the national hero before he was laid to rest. "He was an incredible man," Carl said.

"He was. When you hear the summation of his life, and see the military service, it is easy to see that he was a man of great character," Imar responded.

Carl and Imar walked back toward the car. "I'm really glad you could make it out here for this, Imar."

"I am, too. I wanted to see you and review the progress on Cerebtol. The timing worked for me."

"I am troubled by the comment about Mac's speech issues," Carl said.

"Yes. Me, too. I wonder if it was something other than the virus. Maybe a TIA[20] or something of that nature," Imar said.

"We will never know for sure since he was cremated."

"Unless there is something in the medical records," Imar said.

"I will order them this week and see if they say anything that could explain the episode." Carl opened the door for Imar. "Let's head back to Baltimore and wrap up your review so you can get back to the Alps."

"Let's," Imar said with a smile as he climbed into the car.

Imar continued to gawk at the magnificent oak trees as they drove out of the park. "Beautiful trees," he said. "Beautiful."

---

20 TIA—transient ischemic attack, often referred to as mini stroke, is a transient episode of neurologic dysfunction caused by loss of blood flow.

Removing Obstacles for Approval

June 6

THE HOUSE WAS huge. It was a three-story, Spanish style home overlooking the Sacramento Valley. On a clear day Quinton could easily see the buildings of downtown Sacramento. The stream of cars moving up and down the mountain on highway 50 looked like liquid flowing through veins. Never ending; always moving. He loved the view from his great room. He could sit for hours and just watch the steady stream of cars, mostly headed to South Lake Tahoe for the weekend. It was mesmerizing.

It was Quinton's self-imposed day off. He just got back from Las Vegas after placing twenty-second in another poker tournament. He was beaten by luck. He had his opponent dead to rights until the guy drew a straight flush on the last card. The odds were astronomical. Only one card could beat Quinton's hand and the guy pulled it. The announcers called it a "bad beat." Bad wasn't the word Quinton had in mind. No. It was much worse than that.

But, that's poker.

He was still happy with his prize of $106,316. Yes, very happy. And today was his day off to celebrate and relax. He was enjoying the view and his coffee when his cell phone rang. He automatically reached over to get it, but it wasn't there. He glanced around and found it on the other chair. He grabbed the phone and answered it. "Yeah?"

"Hey, Quint," the young man's voice said.

"Hal. What's cookin'?"

"Thought I'd go to the car show. You game?"

"Yeah. Sounds great. I'll be at your place in about thirty minutes." Quinton tapped the phone to turn it off, slipped it into his pocket and started toward the door. As he passed by the end table he reached down to get his truck keys. They weren't there. He felt the clean table top and froze. He slowly looked at the table. No keys. He thought for a few seconds and walked into the laundry room to check the key rack. There were three sets of keys on four hooks. One was empty. No truck keys. It dawned on him that he forgot where the keys were and he panicked.

Quinton dashed down the hall and into the bedroom. He walked along the dresser top pushing small items out of the way. No keys. He turned and walked into his large walk-in closet. One wall was lined with men's shoes— more than fifty pair. The pants were all on one hangar bar, the shirts on another. He pulled a pair of pants off the rack and searched the pocket. No keys. He tossed them on the floor and pulled a shirt down. No keys. He threw it on the floor on top of the pants and hurried out.

Quinton dashed down the hall to the garage door. He went out to the four car garage. He looked inside the truck. No keys. He slammed his hand against the door and dented it. He didn't care. He was beginning to sweat. He ran his hands through his hair. "Where are they? Where?"

He ran into the bathroom on the first floor, knocking the coat rack over. Coats scattered on the floor. He turned back and checked a coat pocket. No keys. He ran down the hall to the bathroom and looked on the counter as he pushed everything off. No keys. He stood staring into the mirror, sweating. He was breathing heavily. He slowly leaned forward, hung his head and started to slump to his knees. He placed his hands on his hips … and felt the keys.

They were in his pocket.

He pulled the keys out and yelled as he threw them out the door and against the wall. He turned back to the mirror and screamed.

—*—

Owen was sitting at his desk talking on the phone when his secretary peeked around the corner of the door jamb. "Mr. Pitke. Sorry to interrupt."

"Can you hold on for a second," he said into the receiver and pushed a button. "What is it?"

"There's a Quinton Lemolo on the line for you. He says it's extremely urgent, sir."

"Okay. Thanks." The lady turned to leave. "Uh, please close the door behind you." She closed the door.

Owen went back to the first call. "Don. I need to call you back. Something came up. Thanks." He pushed a button. "Owen Pitke here. Quinton? Another episode? ,,, Yes, I am working on it. You were ... what? You what?" Owen leaned forward as he listened to Quinton tell of his latest experience with misplacing his truck keys. "Oh, I wouldn't get all worried over that. It's just a ..."

Quinton told him of the other experiences he recently had; couldn't find his shoes. Got turned around at the airport. Forgot his flight itinerary. It seemed like they were occurring more frequently since his last call.

"Quinton, I think you are under a lot of stress. I'm sure everything is O.K. ... Yes, I called some doctors to run some more tests. ... Yes, we can certainly make sure everything is alright. Yeah. Uh, how about I call you back tomorrow after I make the arrangements? ... Yeah. I have to get the labs ready and the doctors on board. Yeah. Who else did you tell? ... O.K. Good. Yeah, I'll call you right back. Uh-huh. Bye."

Owen placed the receiver in the cradle and thought for a few seconds. He straightened his hair and picked up the phone.

"Don. Owen here. Yeah. I think we have a problem."

—*—

The sports bar was busy. One could watch any of the four main baseball games broadcast that evening on any of the eight screens scattered about the room. People were seated along the bar and at tables throughout the place. Many were sampling appetizers. Most had beers or other drinks. The conversations were loud and filled the room.

Owen Pitke sat in a booth in the corner of the sports bar facing the door, watching. Don Sinclair walked in and brushed off his coat as he looked around and spotted Owen. He looked around the bar one more time to make sure he didn't recognize anyone else as he walked toward the booth. "Hello, Owen," he said quietly.

"Don," Owen said as he straightened his hair. "Thanks for coming."

"Like I had a choice," he whispered. "What's the big deal here, Owen? You know I don't like to meet like this."

"It's Quinton."

"Please, call him 'Q' for now. The less we say, the less the chance of anyone overhearing us, please," he said as he signaled for Owen to lower his voice.

Owen leaned forward. "O.K. 'Q' called me again. That's eight calls in a few weeks."

"Don't panic. It's just a call."

The waitress walked up as Don was speaking. "Hi, guys. Ready to order?"

Neither of the men had looked at the menu, but Don didn't care. "Sure. I'll have the appetizer sampler and a draft, dark ale or something."

"We have amber dark."

"Fine," he said as he smiled.

The waitress looked at Owen. "And you?"

Owen straightened his hair. "I'll have the same."

"I'll be right back with your drinks," she said as she walked away.

"It's more than just the calls, Don. He's ... relapsing or something," Owen said.

"What do you mean 'relapsing.'?"

"He's forgetting stuff."

"Like what?"

"His car keys, appointments, stuff like that."

"Car keys? You're scared because he forgot where his car keys were?" Don chuckled. "Really?"

"Yes, really! His memory was flawless. Now, there're holes in it." Owen looked around. "He's in a panic."

"What do you mean?"

"He's ready to check into the hospital to have some tests run … without my authorization. He wants to go in on his own."

Don looked around the room. "Oh. That's not good. No. Not at all." He thought for a few seconds. "Any other patients showing signs of relapse?"

"No, not that I'm aware of. And any calls like that would come to me."

"Good. Has this guy told anyone about this?"

"No, he said he didn't," Owen replied.

"Okay. We have a lot riding on this, Owen. A lot."

"I know."

Don looked at his folded hands on the table. "If he is suffering some type of relapse, that could shut everything down. We have too many things hanging on this."

"I know. Millions."

"And he was the last of the original test patients?"

"Yes."

The waitress walked up and sat two draft beers on the table. "Here 'ya go, boys. Your food will be right out," she said with a smile.

"Thanks," Don replied.

As the waitress walked away, Owen leaned forward. "What are we going to do?"

"We? Owen, I'll take care of this," Don said.

"You?" Owen straightened his hair. "What are you going to do?"

Don picked up his beer and admired the color and head. "I'm going to make sure we have no obstacles to greatness, my friend."

"How?"

"Leave that to me." Don lifted his glass. "Here's to Lintrovil and Cerebtol, the cure for Alzheimer's," he said as he lifted his glass.

Owen lifted his glass, not knowing what he was toasting, but trusting Don. They clinked mugs as the waitress brought their plates of food. "Here 'ya go, boys. Enjoy," she said as she laid the platters on the table.

"We will," Don said with a smile. "Yes, we will."

June 10

A MAN SAT back, leaning away from the spotting scope and rubbed his eyes. "Still nothing?" a woman asked.

"Yeah. Nothing. He's alone. Has been for hours."

"Good. Looks like we need to move." The woman put a few items into a bag and checked her watch. The man grabbed the scope and shoved it into a duffle bag. They both checked their pistols, holstered them and walked downstairs to the lobby. The man grabbed an apple out of a large bowl as he walked by and stuffed it in his coat pocket. The two headed out to the car.

"We need to get right over there before someone shows up," the woman said.

"Yeah. Only take us a few minutes."

The man started the car and they headed down the street and around the corner. The woman checked her pistol again. It only took them a few minutes until they pulled up to the large Spanish style house. Both of them scanned the area to make sure no other cars were around. "Looks like no one else is here. Let's do this," the woman said.

The woman unbuttoned her top blouse button as they got out of the car, walked up to the door and rang the bell. Quinton opened the door. "Yeah?" he said.

The woman spoke first in a soft, sexy voice. "Quinton Lemolo?"

"Yep." Quinton eyed the beautiful woman, unaware of the man stepping a little closer to him. "Who are …" The man thrust his hand forward hitting Quinton directly in the throat as he placed his foot behind Quinton's leg, knocking the wind out of him and pushing him backwards directly to the ground in a flash. The man and woman stepped through the door. The woman closed the door as the man grabbed for Quinton, still coughing and scooting backwards trying to get away, unable to speak, yell or fight. He was terrified, wild eyed, trying to kick at the stranger grabbing at him as floundered on the floor. He was disoriented and no match for the bigger man. The man grabbed his gun and hit Quinton on the side of the head knocking him out.

"That's just great. I thought you could knock him out with the first punch?" the woman said.

"Yeah. So did I. Rusty, I guess," the man replied.

The pair slipped on a pair of gloves. Then, the man lifted Quinton off the floor. "Chunky, aren't you fella?" he said as he struggled to lift him up and over his shoulder into a fireman's carry. He carried Quinton to the garage as the woman got the car and pulled it up the circular driveway to the front of the garage door. She opened the door and the trunk. The man placed Quinton on the plastic in the trunk of the rental car. The woman looked at Quinton lying unconscious in the trunk on the plastic. "Time to go for a ride, baby," she said as she slammed the trunk lid shut.

The two scanned the front of the driveway. No one. "O.K." the woman said. "Let's wrap this up." The man got the keys to one of Quinton's cars. "This will work."

The woman left first in the rental car. The man waited about two minutes and then left in Quinton's BMW.

The pair met a half mile away and drove through the hills along the upper Consumnes River looking for the right place. The woman came to a turnout that had a trailhead sign. There were no cars there and no traffic. She pulled over and backed up near the trailhead and to the side of the cliff overlooking the river just as the man came up in the BMW. "This is the spot," she said. The man stepped out of the BMW and tossed the keys to her.

She opened the trunk and slipped the keys to Quinton's car into his pants pocket. Then she helped the man lift Quinton out and onto his shoulder. Quinton's head was bleeding slightly from being hit by the gun and dripped on the man's coat. "Damn," the woman said. "We need to clean that up or get rid of it later."

"Great. Another coat bites the dust," the man said. The man started down the trail as Quinton started to come to. He moaned slightly. "Waking up there, fella?" the man said.

Quinton's head was spinning. It throbbed in pain. He tried to focus, but everything was a blur. He tried to speak but his throat hurt so badly. *Am I dead? Where am I?* He tried to lift his arm, but it was in a tight grasp.

The man carried Quinton about fifty yards down the trail as the woman stayed with the cars. When he came to the turn in the trail that overlooked a cliff down to the river, Quinton started to put up a feeble fight as the man heaved him off his shoulder and over the cliff. Quinton let out a slight yell as he sailed through the air looking up in terror at the man. The man could hear the wind knocked out of Quinton's lungs as he hit and bounced off a boulder. "Ouch! Gotta hurt," the man said as he smiled and watched Quinton tumble, roll, and bounce off rocks until he lay still at the bottom of the ravine near the river. "See 'ya," the man said as he saluted Quinton lying lifeless at the bottom of the cliff.

The woman was waiting by the car as the man emerged from the trail. "Ready?"

"Yep. All done," he said as he got into the car. "You check the trunk for blood?"

"Yeah," she answered as she started the car. "We got lucky. There was none."

The pair drove off leaving Quinton Lemolo dead at the bottom of the Eternal Springs trailhead.

—*—

Eight year-old Joey Kruger was excited to go to the ocean. He worked feverishly getting his sand toys, binoculars, hat, coat, sandals, and sun screen just in case it warmed up a little more today and mom would let him take his shirt off and maybe, just maybe, he could get his feet wet. The North Atlantic Coast in June was normally brisk and chilly, but occasionally a warm day surfaced and Joey hoped today was the day. "Mom? I got all of my stuff together like you asked."

Joey could hear his mom struggling down the hall using one crutch and the hand rail along the wall. "O.K. Let me take a look and make sure you have everything." Alice struggled around the corner and into the little boy's room. She could see his pile of items laid out on the bed and floor. "Good. Good. Uh, why are you bringing your wallet?"

"You said we were gonna stop by the store and I could shop for a game."

"No I didn't." Alice started to get perturbed. "The toy store isn't even in the same direction."

Joey became defensive. "Yes, you did, mom."

"Joey. I wouldn't say that. We're going in the opposite direction. I know I ..."

Joey cut her off. "Yes, you did!" He didn't like to make mistakes or have his parents think he was making up stories. He wanted to be honest and smart like his dad, so to have someone imply he was wrong was hurtful and upsetting. "I'm not making it up."

Alice saw her little boy almost in tears and crouched down. "I'm sorry. I guess I just forgot. But we aren't going that way, so maybe tomorrow I can take you." She held out her arm. "Come here."

Joey moved toward his mom "Here. I'm sorry," she said as she wiped a tear away from his cheek. "Forgive me?"

Joey held his head down. "Yeah."

"O.K., then. Let's not have this spoil our trip. Go ahead and put your things into the bag and let's go to the beach!" she said as she clapped her hands bringing excitement back into the journey.

Joey turned and started shoving his things into a bag as Alice turned and started down the hall leaning on her crutch and rail. *It happened again,* she thought. She could feel the sweat breaking out on her forehead. *I need to see the doctor. Should I tell Carl?*

Carl was sitting on the sofa watching the news as Alice rounded the corner. "Look at this!"

On the screen was a young female newscaster interviewing someone. *"Can you tell us what the plans are for this product?"*

The man being interviewed was of Asian descent. He looked to be middle aged, thin and much taller than the woman. He was wearing glasses, had a slight mustache, and was wearing a sport coat. *"After we finish the clinical Phase Two, we will start Phase Three testing."*

"What does that entail?" the interviewer asked.

"Phase Three is a sample group of maybe a thousand test subjects. They are told the implications and risks of the product, which we have found to have no risks, and will be closely monitored to see how the product works over time."

"How much time are we talking about? The world is waiting, Mr. Lee."

"Yes, I know. We expect to have the test completed in less than a year, and the product to market by ... next spring."

"Really? Next spring? You must be pretty excited about this."

"Oh, yes. We are. This could be one of the most significant medicinal discoveries of the century," he said with a big smile.

"Did you hear that? Next spring," Carl said.

Alice stared at the television. "When do you think you will have Cerebtol out?"

"I don't know. We're close. Owen is handling that."

Joey rounded the corner with his bag in hand. "Ready!" he said excitedly.

Alice turned to Joey. "For what?" she asked.

June 14

OWEN PITKE WAS walking toward his office when Carl Kruger walked up behind him, grabbed his arm and spun him around. "Owen, we need to talk. Quinton is dead."

Owen looked surprised. "Dead? Quinton Lemolo?" He could feel the blood flowing out of his face. He thought he was going to pass out.

"Yeah. Let's go into your office."

As the two men walked by Owen's receptionist, the woman started to say something. "Not now," Owen said as he held out his hand and walked past her with Carl a step behind.

"The police found Quinton at the bottom of a cliff a few miles from his home. I got a call from them when they found my card in his wallet."

"Car accident or something?" Owen asked as he straightened his hair.

"Don't know," Carl answered. "They found him near a hiking trail."

"Maybe he fell."

"Maybe. I have a friend in the FBI who lives in Sacramento. He is going to assist with the investigation, with permission from both agencies."

Owen sat upright and straightened his hair. "FBI?"

"Yeah. I called him when I found out. This just doesn't feel right. He was given the green light to look into it for me as long as it was on his own time and didn't interfere with his other cases."

"I didn't know you had a friend in the FBI. I guess that's lucky for us." Owen leaned back in his chair and stared out the window.

"If he was murdered, Jamille will find out why and by whom, I'm sure of that," Carl said.

"Good. Good." Owen could feel his hand starting to shake. He wasn't cut out for this. Not at all.

June 19

THE EL DORADO County Deputy Sheriff pulled up to the Eternal Springs trailhead to see Jamille Larson standing near the edge of the cliff overlooking the river. Jamille had his hands in his pockets and looked like a tourist taking in the scenery. "Agent Larson?"

"Yeah." Jamille pulled out his ID and greeted the woman. "Deputy." They shook hands.

"Please, call me Angela." The young woman looked to be of African descent, but her eyes looked Asian. *Polynesian?* Jamille thought.

"And I'm Jamille." He flashed a big smile, and she returned the favor. "Thanks for meeting me here."

"I have copies of the case here." She handed the documents to Jamille. "I realize you don't have jurisdiction over this, and it's a bit unorthodox, but the Sheriff asked me to come out here to meet you; the FBI. What's this about?"

Jamille smiled. "It's complicated. I'm looking into this because a friend heard this guy died and he was in some sort of medical test, so he asked if I could check it out."

"Well, there isn't much here to look into from what I've heard and seen." Angela pointed to the cliff side. "He fell off the cliff trail while hiking."

"Yeah. That's what I heard from the Sheriff."

"It's a pretty cut and dried case of accidental death, from what we can see," she said. "There was no evidence of foul play."

"Hmmm. You guys have a lot of cases, don't you?" Jamille asked.

"Yeah. Swamped. With these budget cuts we just don't have the manpower to cover the whole county. We even stopped patrols from midnight to four, and that is the busiest time. We don't even respond to property crimes."

"Just file a report and unless something shows up ..."

"Exactly. It's everywhere."

"Yeah." Jamille pulled out some pictures and scanned through them. One caught his eye. "Do you think we can get the digital file on this one to blow it up?" He handed her the picture of Quinton lying on the rocks.

"Sure." The woman made a note. "Why that one?"

Jamille pointed to the shoes. "Those don't look like hiking boots. This guy had money, so I'd expect if he went hiking, he'd have some good boots on for the trek."

"Hmmm." The woman looked at the picture.

"Did you find a water bottle or any supplies on or near the body?"

"Uh, no," she answered.

"Hmmm. Hiking a two mile round trip trail in ninety degree weather and no water. Where did he fall from?" Jamille asked.

"About fifty yards down the trail from what we can see."

The pair walked down the trail to the turn and looked over the edge. "So, from these pictures, it looks like he landed down there, right?"

"Yeah. To the right of that tree. We found blood on the big boulder right there." Angela pointed to the boulder about fifteen feet away from the cliff side.

Jamille looked at the ground around him and slowly leaned forward to look at the edge of the cliff. "Hmm." He took a large rock and rolled it over the edge. The pair watched as a small avalanche formed as the rock rolled down the cliff side, missing the boulder, coming up short by ten feet. The deputy looked at Jamille. "That's interesting," she said.

"Isn't it, though?" Jamille smiled.

"You think he jumped or something?" she asked.

"Yeah. Or something." Jamille turned to face her. Her eyes were captivating. "Do you have time for lunch?"

The woman was taken aback by the request. "Uh, yeah. I guess so. I'm on duty, though."

"So am I, sort of." Jamille said as he smiled. "You lead the way."

—*—

The conference room on the top floor of the Alzheimer's Disease Research Center at Johns Hopkins University was full. A large, wooden table sat in the center of the room surrounded by overstuffed, high-back chairs. Seated around the table was Carl Kruger, several scientists from the Memory Project and Margaret Giddings; Chairman of the Board for Johns Hopkins. She was an elderly woman of great respect with a decisive manner and relished in the leadership role assigned to her by her peers.

Owen Pitke sat at the head of the table opposite her. "Thank you for being here. I wanted to inform you, very simply, that I received word from Don Sinclair at the FDA that ... Cerebtol is going to be approved for the FDA Accelerated Development Review!"

The group let out a hearty applause led by Owen. Ms. Giddings clapped so slightly, holding her enthusiasm back in exchange for leadership. Carl feigned applause. He was still thinking about Quinton.

Accelerated development review is a highly specialized mechanism for speeding the development of drugs that promise significant benefit over existing therapy for serious or life-threatening illnesses for which no therapy exists. The fundamental element of this process is that the manufacturers must continue testing after approval to demonstrate that the drug indeed provides therapeutic benefit to the patient. If not, the FDA can withdraw the product from the market easier than usual.

Owen continued. "This means we will now be able to produce and distribute the product while we run the Phase Three clinical test simultaneously."

"Isn't this just a little premature, Owen?" Carl asked. "I mean, less than two years from Phase One testing and we are taking this to market?"

Owen became instantly perturbed by Carl's question, but restrained himself. He straightened his hair. "Doctor Kruger. It is not my call. If the FDA believes we have satisfied the criteria for testing, and the benefit of the product is worthy of the Accelerated Development Review, then we have satisfied the criteria and are clear to move forward with distribution."

"It is quite unusual to have a product come to market so quickly, Director," Ms. Giddings said with a slight British accent.

"Madam Chairman, it is." Owen cleared his throat. "However, again, I say that the FDA did their due diligence and, if they believe we have satisfied their requirements and have given us the green light, then I believe we should take it."

"Here, here," one of the scientists commented.

"Well, then, what is our next step?" one of the scientists asked.

"Our next step is to be prepared," Owen said. "We need to be ready for media attention, packaging, distribution ..."

"And continued monitoring with the Phase Three testing," Carl added.

Owen scowled at Carl. "Yes, Doctor; continued monitoring. Of course." He smiled, but really felt like slapping Carl for being a wet blanket. Owen straightened his hair, again.

"Well, then," Ms. Giddings said as she cleared her throat. "I suggest, Mr. Director, that you contact the appropriate departments and prepare for ... for ... significant media attention coupled with high demand for the product."

Owen smiled broadly. "We have already prepared media statements, product design and distribution plans and are ready for this to go live. I've had the other departments working on this for more than a month in anticipation of this great moment, Madam Chairman."

"Good," she said. "You are a credit to this program and university, Owen."

She never called him Owen. Never.

He smiled and straightened his hair.

—*—

Owen was walking down the hallway after the meeting when Carl walked up behind him. "Owen, we need to talk about this."

"Carl." Owen turned back to the big man who was cowering over him. "You worry too much. There are no side effects, and, if there are, don't you think we'll see them in the first test group?"

"Maybe. It seems like death follows this thing around. The three people who were the original recipients of this virus ... are dead. I found out the police suspect Lemolo was murdered."

"Murdered? Are you kidding?" Owen sat on a bench in the hallway. There was no one else around as they continued their conversation.

"No, unfortunately, I'm not."

"That's terrible. Do they have a suspect?"

"No. Or motive. Nothing. Nothing solid."

"Then, why do they suspect murder?"

"Circumstances."

"What?"

"I told you earlier that they found him at the bottom of a cliff near a hiking trail."

"Yeah."

"Well, he wasn't wearing hiking shoes, had no water bottle, nothing to indicate he was hiking," Carl replied.

"Maybe he committed suicide, jumped or something."

"Possible. But there was one clue that leads them to suspect otherwise."

Owen was very interested. "What was that?"

"He registered for a poker tournament for the next weekend in Atlantic City and purchased the airline tickets that day."

"Hmmm. I guess he could have gotten some bad news or something," Owen said.

"Maybe. Except he purchased the tickets online about thirty minutes before he went 'hiking.' Doesn't fit."

"Yeah. That is odd." Owen could feel the sweat starting to break out as he straightened his hair.

"My friend has a theory, though," Carl said.

"Theory?"

"Yeah. He thinks it might be a mob hit."

Owen perked up. "Mob? As in organized crime?" he asked.

"Yeah. He thinks Quinton may have set someone off playing blackjack and winning so much. The casinos don't like to lose, you know."

Owen could feel a huge relief being lifted from his shoulders. "Yeah. I'll bet that's what it is. A mob hit," he said as he stood. "Can you keep me posted of the investigation? I'd like to know what happened to him, too."

"Sure. If I find out anything, I'll let you know."

Owen started to turn to walk away as Carl stopped him. "One more thing, Owen. With Quinton gone, and the original three recipients, we have no way of knowing if there are any long term side effects of this thing. Death seems to follow this virus, like 19Q. I don't like it."

"Well, you need to learn to like it." Owen took a deep breath and faced Carl. With the FBI believing a mob hit was cause for Quinton's death, he felt assured he could take control and move forward without interference from anyone, especially Carl. "It's going live in a few days. Like I said in the meeting, I've been arranging a product release campaign. We're announcing Monday, July third that we have a cure for Alzheimer's Disease that is going to be distributed starting July fourth; Independence Day. The theme is 'Freedom from Alzheimer's.' Like it?"

"Clever. But I still feel it's premature."

"Well, be that as it may, we're still moving forward. You heard the board. This is our opportunity for the world to see Johns Hopkins as the premier medical research university in the world. We'll be famous!" Owen said as he slapped Carl on the shoulder.

"No doubt," Carl said as he turned and faced the window, feeling defeated. "How's it going to be distributed?"

"Good question. Recipients will complete applications with the person's primary care physician …"

"Applications?" Carl asked.

"Yeah. The person must be diagnosed with Alzheimer's or at extreme risk primarily from genetic factors, and obtain their doctor's approval. Then they will apply for the inoculation and be scheduled sometime thereafter. That's my idea so far, and I have William working on the specifics."

"I have to say, Owen, it is amazing that a single shot can cure Alzheimer's; no continuing prescriptions, no continuing costs, just BAM! Done."

Owen ignored the comment. "You know we've been anticipating this. Timore prepared 500,000 doses, at $5,000 per dose …"

"Five thousand?" Carl was shocked. "Are you serious?"

"I am," Owen said with a smile. He relished the moment.

"That's ..." Carl thought for a second. "... that's 2.5 billion dollars on the first pass!"

Owen smiled. "People will pay that at the drop of a hat. I mean, it can cost a thousand bucks to have your colon inspected. Five thousand to cure Alzheimer's is cheap! There's more than two hundred billion spent each year just on care for Alzheimer's patients!" Owen thought for a second. "We may have to raise the price."

"Raise it?" Carl asked.

"Now you know why that company ... what's their name?"

"Antrole Pharmaceuticals?"

"Yeah, Antrole ... their stock soared when they announced they were working on a cure for AD."

Carl was still thinking. "If we look at the 36 million people worldwide, at a fifty percent penetration, we'd be looking at ... at." Carl pulled out his phone and started punching numbers. He stopped and lowered the phone in disbelief.

"A lot of money, Carl. A hell of a lot of money."

"Almost one hundred billion dollars!"

"And at five thousand dollars, that's a steal." Owen was smiling broadly, thinking about the riches and fame soon to be realized.

"This is all about the money, isn't it, Owen?"

Owen feigned being offended. "Of course not, Carl. It's about helping people, being at the forefront, adventure, a new frontier ..." He turned, looked out the window and spread his arms wide, and continued. "... the University ..."

"Yeah. The University," Carl said in disgust, as though he finally realized what Owen's real motives were. He turned and stormed down the hallway.

Owen watched Carl's reflection in the window as he left. "And, yes, the money," he whispered.

June 20

FDA NEWS RELEASE

For Immediate Release: June 20, 2017
Media Inquiries: Jennifer Sands, 301-555-4666,
jennifer.sands@fda.hhs.gov
Consumer Inquiries: 888-INFO-FDA

FDA approves Cerebtol to prevent dementia in people with either Alzheimer's Disease or Early Onset Alzheimer's Disease

The U.S. Food and Drug Administration today approved the dementia preventing drug Cerebtol (cereboxalan) to reduce the risk of dementia in people who have dementia related to Alzheimer's Disease or Early Onset Alzheimer's Disease.

In Alzheimer's Disease, as in other types of dementia, increasing numbers of nerve cells deteriorate and die. A healthy adult brain has 100 billion nerve cells, or neurons, with long branching extensions connected at 100 trillion points. At these connections, called synapses, information flows in tiny chemical pulses released by one neuron and taken up by the receiving cell. Different strengths and patterns of signals move constantly through the brain's circuits, creating the cellular basis of memories, thoughts and skills.

In Alzheimer's Disease, information transfer at the synapses begins to fail, the number of synapses declines and eventually cells die. Brains with advanced Alzheimer's show dramatic shrinkage from cell loss and widespread debris from dead and dying neurons.

"The breakdown of neuronic activity is the primary cause of the disease," said Owen Pitke, Director of the Alzheimer's Disease Research Center at Johns Hopkins University. "This approval gives doctors and patients a tool, a cure, if you will, for a condition that we consider to be a scourge on our world."

Alzheimer's Disease can affect different people in different ways, but the most common symptom pattern begins with gradually worsening difficulty in remembering new information. This is because disruption of brain cells usually begins in regions involved in forming new memories. As damage spreads, individuals experience other difficulties.

In advanced Alzheimer's, people need help with bathing, dressing, toileting, eating and other daily activities. Those in the final stages of the disease lose their ability to communicate, fail to recognize loved ones and become bed-bound and reliant on 24/7 care. The inability to move around in late-stage Alzheimer's Disease can make a person more vulnerable to infections, including pneumonia (infection of the lungs). Alzheimer's Disease is ultimately fatal, and Alzheimer-related pneumonia is often the cause.

The safety and efficacy of Cerebtol were evaluated in a clinical trial with more than 400 patients. In the trial, Cerebtol was effective in 100% of the patients with significant memory improvements in more than 84% of the patients with no side effects. All patients experienced some memory improvements.

Cerebtol is taken one time as an injection and requires no further administration of the medicine.

There was no adverse event reported by any patients treated.

Cerebtol can only be administered by a qualified physician.

Cerebtol is marketed in the United States by Timore Pharmaceuticals, Inc., a recipient of and partner with the Howard Foundation of Johns Hopkins University.

On June 20, 2017, the FDA approved Accelerated Development Review of Cerebtol to reduce and reverse the effects of Alzheimer's Disease.

Freedom

July 3

GOLDEN POND MEMORY Care in Elk Grove, California was busy; busy like they had never seen before. Two news cameras were outside the building getting shots of the exterior, the walkways lined with flowers and the wetlands nearby with residents being pushed along the boardwalk. Two cameras were inside the building filming residents shuffling along, interviewing employees and interviewing Melanie Grimes, the administrator.

"Yes, this is a very exciting time," she said. Her poise made her look younger than her years. "Most of our residents qualify for the new inoculation of Cerebtol. Many have guardians and are unable to communicate on their own behalf. If this really works like they say it does, then we can expect great results."

The young Middle-Eastern lady interviewing Melanie pulled the microphone away and held it close. "Is there any concern that you will be losing patients and, therefore, losing income for the company?"

"You mean residents. No. We will still provide memory care to those residents that have brain injuries or mental illness. We also have a business plan to move aggressively into rehabilitation and respite care if this product works. We know there is a growing market as the boomer generation ages and rehabilitation from surgeries and injuries is big. We expect that to replace the portion of our memory care services that could leave with alternative services."

"How many patients, I mean residents, will be inoculated tomorrow?"

"From what we have seen, about forty-six. That is just a third of the residents living here."

"I'll bet you are excited."

"That is an understatement. I would say 'awestruck' is more appropriate."

The reporter turned and faced the camera. "And so begins a new era where forty-six people here at Golden Pond Memory Care will receive the new Cerebtol inoculation to combat Alzheimer's Disease. This is just a drop of the projected one million people that will receive the inoculation over the next few months starting tomorrow, Independence Day. Their position, as reported by family and friends, is that they have nothing to lose, and everything to gain. I'm Natasha Aziz from KCRA Three news."

Memory Care facilities all around the country were experiencing the same types of interviews. Tens of thousands of them representing several million people suffering from Alzheimer's and other forms of dementia now have hope.

—*—

Doctor Eldon Northridge was planning on closing his office the Monday before the Fourth of July. But with the news of Cerebtol inoculations being available July Fourth, he had no choice but to open Sunday and work through the holiday. His patients or their representatives made hundreds of calls the prior week and scheduled times to meet with him to schedule an injection. They were excited to try something to stop their suffering.

He would need to meet each patient, review their file and diagnosis, conduct a simple physical exam and perform the injection. They, or their representative, also needed to sign a liability waiver. The waiver indicated they knew the medicine was under the Advanced Distribution Review program of the FDA and would waive any liability associated with risks from taking the medicine. It was a preventative clause for the FDA, the manufacturer and the doctor. No one wanted to be sued if something went wrong. Just in case.

The first and second phase recipients were doing exceptionally well. There was no indication of any adverse reactions from the medicine. It has been more than two years since the Phase One test group received their injections and they have excelled. Phase Two recipients received their inoculations more than a year ago, and hundreds have been discharged from memory care homes. The results were more than encouraging; they were considered a miracle by many.

Including Doctor Northridge. He was excited to be a part of history. Anything to cure this dreaded disease.

—*—

The young, female gerontologist was examining the elderly woman sitting on the side of an examination bed. The elderly lady was drooling slightly. She stared vacantly at the picture hanging on the wall across from her. A little drool was escaping the side of her mouth and fell onto her gown. A middle-aged woman standing next to her wiped the drool away. "The doctor's going to see if you can have the shot, mom," the woman said as she smiled at the older woman.

The elderly woman never changed expression. The doctor listened to her breathing and heartbeat. She checked her pulse, blood pressure, glanced through some charts and made some notes in the computer. "O.K. I think Laura will benefit from this new medicine. She's in pretty good health, all things considered."

"Great. Hear that, mom?"

No expression.

The doctor continued. "All we need to do is have you sign this waiver for her and we will schedule the injection." The younger woman took the waiver and signed it without reading it. "Aren't you going to read it?"

The younger woman looked at the doctor. "My mom has been suffering with this disease for years. Look at her. She isn't even there. That's not my mom, doctor. That's her shell." The woman looked at her mom sitting helplessly, drooling slightly with a vacant stare. "I want my mom back ... at any cost." She handed the waiver to the doctor, who took it and filed it in the folder.

July 4

THE STORE WAS busy. People were buying groceries for their Fourth of July gathering. Many people had bags of charcoal for their barbeques; others had sandwich makings, or hot dogs and ribs. It was a barnyard slaughter of chicken, pork and beef headed to the cash registers. Carl was doing the shopping for Alice while she stayed home to start the meal. They planned on having several friends over for the day, play some games and after sunset let the kids shoot off some fireworks, just like most every American family would do that day.

As Carl stood in the checkout line with his basket full of groceries, he noticed a camera crew pulling up outside the store. He saw the cameraman and the young male reporter walk across the lot toward the store and start shooting.

Carl paid for his groceries and started out the door only to be stopped by the reporter and a small group of people gathering around. The reporter held the microphone out as he approached. "Are you Doctor Kruger?"

Carl was surprised by the question. "Uh, yes, I am. May I help you?"

"I'm Jeremy Lu from KCRA News. I'd like to ask you a few questions about Cerebtol, if I may."

"Well, normally I'd do these interviews in my office, not in a parking lot of my grocery store."

"I understand, but we wanted to get some comments from the public about this remarkable drug, Cerebtol, and their reactions, too."

"I see." Carl looked at the camera. "Are we rolling?"

"Yes, but we are planning on editing this for the news tonight. This is not a live broadcast."

"Oh. How did you know I was here?"

The reporter ignored the question and asked his own. "Were you one of the key scientists that developed the drug, Cerebtol?"

"Yes, along with Doctor Imar Spaan and others."

"Can you tell us a little about how it works?"

Carl explained some of the basic functions of how the drug worked to the wonderment of the growing crowd of viewers standing nearby. One of the ladies in the crowd yelled, "You're a Godsend." Another yelled, "Thank you for saving my dad."

When Carl finished the explanation, he started to walk away. "I'm sorry, but I have some groceries here that I need to get home. If you want another interview, please contact my office." As he started toward his car a young man held out a paper and pen and asked him for his autograph. "You want my autograph?" Carl asked.

The young man was smiling. "Yeah. You're going to be famous like Jonas Salk, or Louis Pasteur. You bet I want it."

Carl signed the paper and handed it back to the man. Other people started to hold out papers and pens. "I'm sorry, folks. Maybe next time," he said as he put the groceries in the car, got in and drove through the crowd. The camera followed him the entire time.

"That was Doctor Carl Kruger; the scientist who discovered the cure for Alzheimer's. Today, July Fourth, tens of thousands of people all over the country will be receiving an inoculation of Cerebtol, the miracle drug from Johns Hopkins University and they can thank Doctor Carl Kruger for discovering it and giving them hope. This is Jeremy Lu from Channel Three News."

Break

September 21

CARL WAS DRIVING the BMW through the mountains to Chestertown, Maryland for a get-away weekend. He didn't have to be back to the office anytime soon, so he decided to take a few days off and spend a romantic weekend with Alice. Joey was staying at his cousin's house, so Carl had no concerns about his son having a great time while he and Alice took off to have fun of their own. They needed a break. The past few months had been grueling; news interviews, media, projects; there was so much attention to the creation and distribution of Cerebtol that Carl was overwhelmed. He guessed Alice felt the same way because she seemed to be downcast quite a bit, so he thought a trip through the mountains on a fall day and a weekend getaway at the Brampton Bed and Breakfast would be just the medicine.

Carl pulled up to the front of the 1860's plantation turned B&B and parked under a giant oak tree. He opened the door for Alice and held her hand as she struggled out of the car and onto one crutch. She held onto Carl's arm as she usually does and the two walked up the stone steps to the grand entry framed by four large, white columns. She could smell the lavender as they approached the large, wooden entry door. "Carl, this is beautiful."

They checked in and got the keys to the cottage around the corner. Without elevators, Alice would never be able to manage the stairs. Besides,

Carl wanted some privacy so they could catch up on life, romance and marriage.

The cottage was just around the corner from the main house. There was a stone walkway lined with roses and mums that twisted across the grounds and under the large birch trees. The lawn was manicured. Carl wished he had brought his clubs so he could practice his short-iron shots. Probably best he didn't.

The cottage was small, white, and nestled between a babbling brook and a large willow tree. A peacock strutted across the lawn in front of them. As they entered the cottage, Alice let out a deep sigh. "This is just magnificent, Carl."

"It is," he replied as he tossed his jacket onto the bed. "O.K., I'll get the bags from the car while you settle in." He turned to go to the car when Alice grabbed his arm to stop him.

"I don't have a bag," she said.

"What? What do you mean?" he asked.

Alice became downcast, almost to tears. "I ... I didn't bring a bag, Carl." Tears formed in her eyes as she started to cry.

"Alice?" He helped her sit on the edge of the bed as she continued to cry. "Honey. Will you tell me what's going on? Why are you crying?"

Alice calmed down and, with tears in her eyes, said, "I forgot the bag."

"Oh, that's O.K. It's nothing to get so upset about. I forget stuff ..."

"Carl!" Alice yelled. "I forgot, again! I'm forgetting stuff all the time!" Alice's sadness almost turned to anger that Carl didn't understand.

"What? What do you mean?"

Alice fought back tears as she told Carl of the numerous times she forgot something; the time she forgot to pick up Joey and found herself driving on the highway toward downtown. And the time she forgot to take Joey to his friend's birthday party after she and Joey got in the car and started driving there. The lost keys, purse, phone, putting different colored socks on, appointments, her teeth. There were so many things. And they were building. "Carl, I'm scared."

"Oh baby." Carl pulled Alice toward him and cradled her head against his chest. "It's O.K. It's going to be alright."

"Carl," she said as she sniffed back her tears. "I'm really scared. I don't want to lose my mind like my dad. That was terrible."

"Honey, it will be alright. We'll go see the doctor and see what's going on. It could be vitamins, or hypoglycemia, or …"

"Carl! It's none of that!" She pulled away from him, angered that he was not taking this as serious as she was. "You don't understand."

"Then tell me," Carl said pointedly.

"It's getting worse! I've been tracking it. The frequencies. Occurrences have increased almost five-fold in two years."

"You sound like a scientist."

"I *am* a scientist!" Alice shot back. "I just happen to be working as a mother right now."

"Honey. Calm down." Carl put his arms around her shoulders. "I can see you really are scared here. I'm sorry. I didn't mean to take it lightly. I was just trying to reassure you."

"This is serious, I can feel it. I can tell. It's getting worse and I don't know what to do and I'm scared."

"Do you have an appointment with a neurologist?"

"Not yet. I wanted to talk with you first."

"O.K. Well, we can see one next week when we get back. Let them run some tests and see what the diagnosis is. Then we can figure out where to go from there."

Alice rested against Carl, feeling the rise and fall of his breathing and his big arms around her. She felt safe. She was relaxed and protected. "Uh huh."

"Let's just relax this weekend and enjoy one day at a time."

"O.K."

Alice and Carl sat on the edge of the bed, Carl holding Alice and gently rubbing her back. "Carl?"

"Uh huh."

"Don't leave me."

Carl was a little surprised. "I'm not going anywhere, baby."

"If this is bad, please don't leave me."

Carl pulled Alice close to him. "You don't have to worry about that because it's not going to get any worse."

Alice sat up straight. "You don't know that. No one does."

Carl paused for a second and looked Alice in the eyes. He could see her fear. "Alice. I will never leave you. Never."

That was what she wanted to hear. She laid her head against his chest and soon fell asleep as he rubbed her back.

Carl could feel the fear starting to rise up in him as he held Alice.

—*—

The response to Cerebtol was astounding. More than 100,000 people were inoculated in less than a week; nearly a million in less than three months. The first round of inoculations was very controlled. Only patients diagnosed with advanced Alzheimer's Disease or Early Onset Alzheimer's qualified for the inoculation. The people with Early Onset Alzheimer's only made one percent of the test group. They were highly functional but at extreme risk of contracting the disease at an early age. Their inoculation was strictly preventative. Since they were capable of making their own decisions, they only needed to prove the existence of significant risk factors in their family to obtain the inoculation. No one expected any immediate results of the inoculations. They only hoped Alzheimer's Disease would never manifest itself, and they could continue to live normal, unabated lives.

The future of the patients with Advanced Alzheimer's, however, had very little to offer and the quality of their lives was miserable. They were considered a test group that could afford the risk. They existed; not lived. Many of those had guardians in place to make medical decisions for them. Others had family members on record as the decision maker. In any event, many were unable to make their own choices because of the severe mental deterioration, so someone else had to decide for them.

Nearly all of the recipients were housed in some type of a locked facility. Very few lived at home with twenty-four hour in-home care. The cost was prohibitive running above $15,000 a month. The facilities were usually crowded with two to three people sharing a room. The rooms were full and most facilities had waiting lists. It was hard to believe that, even in these crowded conditions, patients were paying $7,000 or more per month for memory care.

Each patient that received the inoculation was closely watched, at least as closely as resources would allow. Even though just a couple of months had passed, many were showing improved memories. They were able to feed themselves, or toilet without assistance. Many were able to converse again at a basic level. Word was spreading that maybe, just maybe, Cerebtol *was* the answer. It was the miracle drug that the world had been waiting for.

And the media was eating it up, story after story of its success, and all of it was attributed to Carl Kruger and Imar Spaan from Johns Hopkins University.

A New Lease

December 22

THE SNOW WAS falling steadily. Carl Kruger was sitting at his desk at home wrapping up the affairs of the week, getting ready to take a couple of weeks off for Christmas vacation. He was excited to spend time with Alice and Joey and just relax around the house. He planned to take them to New York for a few days to let Joey see the skyscrapers and the Statue of Liberty. He even got reservations to The Clown on Broadway. But the most important place to visit for Carl was Ground Zero. He lost several friends in that disaster, and going to Ground Zero was a way that he could pay respects and reflect on how the world had changed. He ventured there every year after that day. It was a way for him to heal personally.

Carl could hear Alice's crutches coming closer as she approached the study. He turned to greet her before she arrived at the door. "Honey," Alice said as she walked into the study to see Carl smiling at her. "Isn't it time to stop? It's almost six."

"It is," Carl said as he stood and pushed the chair under the desk. Carl and Alice walked out to the living room and sat on the sofa. "I'd like to see the weather forecast for the weekend. I bet it snows the whole time we are in New York," he said as he turned on the television.

The newscaster was standing by a lighted Christmas tree outside a beautifully decorated building. Some people were in the background loading some bags and a television into an SUV. An elderly man was getting into the passenger side with the use of a walker. A younger couple were assisting him and getting the stuff into the car. *"… and it's going to be a great Christmas for many people thanks to the miracle drug, Cerebtol. It's changing hundreds of thousands of lives already, like Mr. Evans' here. Once a confused, elderly man unable to feed himself. Now, going home to spend Christmas with family that he fondly remembers."*

The screen changed to a close-up of the elderly man with a big smile on his face looking into the camera. *"Merry Christmas,"* he said as he waved his gloved hand, his breath creating a small cloud.

"I'm Neil Slokum from Channel Three News. Natasha?"

"Thanks, Neil. What a great story. It looks like it's going to be a very Merry Christmas to hundreds of thousands of people all over the country thanks to this new drug developed by Johns Hopkins University through Timore Pharmaceuticals." A picture of Owen Pitke appeared on the screen.

"Look." Carl said as he leaned forward.

"The brains behind the brain drug is Owen Pitke, Director of the Alzheimer's Research Center at Johns Hopkins University here, in Baltimore."

"Just like Owen," Carl said, staring at the image of Owen with a bad comb over.

"Doctor Pitke was instrumental with coordinating a team to research and develop this drug over the past several years. Key team members included a local doctor, Carl Kruger, and a Doctor Imaar Spaan, a noted microbiologist from Vienna." A picture of the two flashed on the screen. Imar's hair was wild, as it always was. *"It looks like a lot of people will be thankful to these folks for their new lease on life."*

"Well, Mr. Celebrity," Alice said as the picture disappeared. "You can probably start expecting more phone calls and interviews about this."

"I guess," Carl replied. "I've had enough already. I'd rather have the attention go to Imar or Owen. Owen loves it, anyway. I don't, not really."

"That's what I love about you; your humility," Alice said as she wrapped her arm around Carl's.

Joey ran into the room and jumped onto the sofa. "Let's watch a movie," he said.

"O.K. How about our Christmas tradition, The Santa Clause?" Carl asked.

"Yeah." Joey jumped up to put a CD into the player.

"I'll make some popcorn," Alice said as Joey took her place on the sofa next to his dad.

The three of them enjoyed the warm fire, popcorn and a movie. Carl thought how perfect his life was.

Year Five—2018

Diagnosis

March 16

THE NURSE STEPPED into the lobby holding a file in her hand. "Ms. Kruger?"

"Yes." Alice put the out-dated magazine down and lifted herself to her crutches. "Right here," she said as she struggled toward the door.

"How are you, today?"

"Fine. Just doing some 'soccer mom' errands."

The ladies chatted as they went down the hall, weighed Alice and entered a small room. The nurse went through her routine, checked blood pressure, temperature, heartbeat. Everything was fine. The nurse entered some data on the monitor and wrote some notes on a pad. "The doctor will be right with you," she said as she exited the room, closing the door behind her.

This gave Alice time to think.

I know if Carl finds out, he'll be mad. Well, maybe he won't. He's pretty understanding. Like that time I caught the stove on fire when I forgot the egg was frying. It could have been a disaster. I just forgot. It was important, the phone call. I hadn't heard from my sister in a long time. She was having some struggles, too. How old is she? Was it 1964 or 65 when she was born?

"Hello, Alice," the doctor said interrupting her thought. "How are you today?" he said with a smile. He was an older man, distinguished in appearance and fit.

"O.K. Well, sort of. I'm ... I'm forgetting quite a bit lately. This is starting to interrupt my life, doctor."

"Alice, you are a scientist," he replied.

"Was, doctor. Was. I haven't been active in my field for five years, almost."

"Once a scientist, always a scientist, Alice. You still have the thought process of analyzing and evaluating and curiosity. You know something is going on, as do I." The doctor flipped some notes in the chart file. "Your tests came back positive, Alice. You have Early Onset Alzheimer's Disease."

Alice became downcast and stared at the floor. She could feel the fear well up in her stomach. "I know what that means, doctor. I'm already experiencing some accelerated memory loss."

"Sorry to hear that, Alice. But there is hope. And in large part thanks to your husband."

"I know. But I haven't told him."

"You ... you haven't? Alice, I'm surprised. He has discovered the cure for your disease and you haven't told him?"

Alice struggled to find the words to express herself. "No. I ... if it didn't work I ... I didn't want him to feel like he let me down or was at fault or something. I just ... I ..."

"Well, it is your choice. But it does work, Alice. The FDA has approved it, people are healing; it works." The doctor looked through the notes again. "Your medical condition is your business, Alice, and I respect that. As your doctor, I suggest you take the inoculation. Otherwise, your memory will continue to deteriorate significantly, as you are aware."

"I know." Alice paused. "What if I wait a year or two?"

"You could do that, but I'm not sure what purpose that would serve."

"It would give me more time to ... think about it, I guess. I don't know."

"You can do that. But your memory will continue to deteriorate, Alice. Early Onset Alzheimer's is significant. It accelerates the memory loss and some of your motor skills. You'll probably start getting pretty frustrated about it, and that will affect your quality of life, relationships …"

"It is already, doctor." Alice paused. She took a deep, cleansing breath. "I'd like to have the inoculation, but I'm not going to tell Carl. Not yet. Maybe later. I don't know."

"That's your call." The doctor closed the file. He pulled out an information sheet and liability waiver. "Here's the information on the product and the liability waiver. I need to have you take this home, review it, and, if you still agree, come back in for the inoculation."

"Why can't we just do it today?"

"We could, but in your case I want you to make sure you want it, Alice. No second thoughts here. Not with you. Not with Carl involved. This is different. O.K.?"

"Sure. Can we go ahead and schedule it? I have a pretty tight schedule running Joey all over town, so I'd like to get it on the calendar."

"Sure. I'm going to schedule this personally. I think it's better to just have the two of us involved and no one on my staff, just to make sure. I don't think they would say anything, in fact, I know they wouldn't, but I really do not want to take any risks here because of the situation." The doctor turned to the computer. "How about …?"

"Monday," Alice said.

The doctor turned and looked at her. "Monday?"

"Yeah. Monday. That gives me the weekend to think about it one more time."

"Huh." He turned back to the screen. "Good. Monday it is. How about three?"

"Sure. I'll be here with the signed releases."

The doctor turned back to Alice and shook her hand. "Then Monday it is."

The Notice

June 15

"CARL!" OWEN BURST into the office with a document in his hand. "Did you see this?" He was excited and smiling widely as he flapped the paper around.

"What? See what?" Carl reached for the paper.

"The Nobel Prize, Carl. The Nobel!"

"What?" Carl asked as he scanned the document.

"You and Imar are being nominated for the Nobel Prize in medicine!"

Carl lowered the document and looked at Owen. "This is real!"

"Yes, it is!" Owen slapped Carl on the shoulder. "The Nobel, Carl!"

Carl got a huge smile as he looked at the paper again. "Owen, I never imagined I would ever see something like this. I mean, just to be nominated ..."

"This is amazing!" Owen straightened his hair and paced around the room. "This is only the third time a doctor or scientist from Johns Hopkins has been nominated for the Nobel. This is going to put us on the radar as the premier medical research institution in the world!"

"I remember the work Doctor Carol Greider did on chromosomes, telomeres and the enzyme telomerase. I used much of that research when I was at ICE. And now, to think ..."

"Carl." Owen stopped and paused. "Do you realize we are making history?"

"Yeah. It's surreal." Carl grabbed the phone. "I need to tell Alice." He quickly hung up the phone. "No, I'll tell her in person." Carl grabbed the phone again. "I need to tell Imar. Does he know?"

"No, not that I know. I haven't been in touch with him since he left for Austria."

Carl quickly dialed Imar. "He is going to be floored. What time is it there? It doesn't matter anyway."

Owen was clapping his hands, applauding the moment as Imar answered the phone. Carl switched to the speaker phone. Imar's voice was groggy and raspy. "Hello?"

"Imar! It's Carl!"

"And Owen."

"What? What time is it? Why are you calling me? Is something …?"

"Imar," Carl interrupted. "We've been nominated for the Nobel!" There was silence on the line. "Imar? Did you hear me? I said …"

"I heard, I heard," came the raspy voice. "The Nobel, huh?"

"Yes, Imar. The Nobel. You and me."

"That is quite astounding, my friend. Congratulations."

"And to you, you wild haired genius!" Carl replied as he laughed. "We'd love to see you again. Any travel plans?"

"Well, I suppose I will be going to Sweden in December for the ceremonies."

Carl chuckled. "I know that. Do you think you will come out here before then? We miss you."

Imar cleared his throat. "Oh, I don't know. I am kind of enjoying my semi-retirement. I will have to give it some thought. It's a long way."

"Well, if you do, you can stay with me and Alice for as long as you like this time. We have room and would love to have you."

"I will consider. Now, may I go back to sleep?"

Owen and Carl laughed. "Sure. Nighty night," Carl said as he hung up. "I guess he wasn't too excited, being his second Nobel."

Owen started to pace the room as he spoke. "I expect we will have a full media blitz here. We need to get this out to everyone to see what JH is doing," he said. "Media, marketing, the local channels. Why, the national channels. I need to have marketing contact Good Morning America. We'll need to do some speaking engagements at …"

"Hold it. Hold it. I'm not sure I'm ready to start traveling the country speaking on this, Owen."

"Country?" Owen moved closer to Carl. "I'm talking the world here, Carl."

"What?"

"This is it. This is the biggest discovery this century. You and I need to represent the University."

"I've already done enough interviews to last a lifetime. I have a family, Owen."

"They can go with you."

"Joey has school. I can't just pull him out."

"I'm sure we can plan several trips that only last a few days. The longer ones, likely international, we'll plan now, during the summer while he is out of school. You can use it to expand his knowledge of geography and culture."

Carl let the thought sink in. "Maybe Imar can do some of the European presentations."

"There 'ya go," Owen said as he straightened his hair. "I can do some as well. Teamwork. That's what this is about. Teamwork."

"I'll have to clear this with Alice."

"Good. I'll start working on itineraries. You'll need to be ready for the media. I expect they'll come knocking at your door again, pretty soon."

"Yeah. I guess they will."

The Society of Neuroscience

July 4

THREE WEEKS HAD passed since Carl Kruger was notified that he and Imar Spaan were nominated for the Nobel Prize in Physiology and Medicine. It was an honor that all doctors and scientists strive for, yet few obtain. The publicity that follows such an achievement catapults the nominee to new levels of stardom in their respective fields. The number of interviews with a variety of media escalated immediately following the announcement. Paparazzi were following him and sometimes his family. Fame had its price.

Such was the case with Carl Kruger.

His calendar was quickly filling in with speaking engagements around the country. From New York to San Francisco, Doctor Kruger was known as the man who cured Alzheimer's Disease. Even though the FDA still had the JC2 virus in an Accelerated Development Review, the product was flawless and was viewed as perfect. Everyone that was inoculated showed significant memory improvement, and the younger recipients had an increase in intelligence. People with progressive Early Onset Alzheimer's saw immediate change. There were no side effects. None. Just memory improvement.

The world of memory care and Alzheimer's Disease was rapidly changing. The success of Cerebtol, marketed by Johns Hopkins through Timore Pharmaceuticals, and Lintrovil, sold by Antrole Pharmaceuticals, was remarkable. Memory care facilities discharged hundreds of thousands of residents who eventually moved to a lower level of care, leaving hundreds of thousands of rooms empty. They began targeting new markets in the "active retirement" population. Some facilities went bankrupt, or were sold to become apartments or motels by combining rooms and gutting common areas into restaurants and lobbies.

Memory Care was quickly fading away; in just one year. Only brain injured and some dementia based memory loss patients were still housed. New Alzheimer's Disease patients were few and far between. Alzheimer's Disease was quickly going the way of polio, measles, and other once prevalent diseases into the land of obscurity.

The world thanked Carl Kruger and Imar Spaan for the discovery.

The anniversary of the announcement and initial inoculation of Cerebtol was today, July 4th. Carl was scheduled to speak at the Society for Neuroscience in San Francisco, California to celebrate the initial inoculations of Cerebtol. The news of him being nominated for the Nobel instantly shot him to the top of the list of desired speakers at this conference. The program was developed almost a year before with Carl as a speaker on the new drug approved by the FDA. The topic was going to be about the discovery and development of Cerebtol. It was estimated nearly 50,000 doctors, scientists, and educators would be present at the conference to hear this new medical celebrity speak.

Carl, Alice and Joey flew to the "City by the Bay" and planned an entire week of vacation mixed with Carl's presentation. Alice and Joey planned to spend the afternoon at the zoo when Carl's presentation was scheduled. It didn't interest Alice anymore. She heard it before, and now it was getting kind of boring. She could recite the entire speech verbatim. She wanted Carl to change it a little, but he wasn't comfortable deviating from the facts. So, the zoo was much more appealing to her even though she had to contend with the crutches.

The stage of the massive Louise M. Davies Symphony Hall in San Francisco was arranged with four chairs, a lectern and a small speaker. Behind the lectern were two large screens positioned so each person in the three-thousand seat sweeping auditorium could see the presentation. Two cameras were zeroed in on the lectern and stage so the remaining 45,000 attendees in various locales could watch the broadcast live. Carl sat center stage with the Chairman of the Society, another noted doctor of neuroscience, and Owen Pitke.

"I am so glad you were able to make it here for this," Carl said. "It helps to have another person to go to with questions."

The Chairman of the Society leaned toward Carl and whispered, "Shhh. They can hear you. The acoustics are phenomenal here," he said. The elderly man leaned back in his chair and watched the speaker.

Owen straightened his hair and smiled.

The doctor at the lectern was wrapping up her short presentation on steroids and plasticity and their effects on neuroendocrine processes. The audience applauded at the right time as she finished her presentation. There were no questions. Everyone was anxious to hear about Cerebtol from the Master.

The Chairman approached the microphone and cleared his throat. The speakers were very small but of extremely high quality. They provided just enough amplification to enable any speaker's voice to be heard throughout the auditorium with ease and clarity. "Thank you Doctor Miles. And now, we have Doctors Carl Kruger and Owen Pitke from the Johns Hopkins Alzheimer's Disease Research Center to share their insights on the development of Cerebtol and Alzheimer's Disease." The elderly man read the accomplishments and certifications of the two noted speakers. When he finished, he stepped back and Carl rose to thunderous applause.

The auditorium quickly quieted down, and the audience listened intently to every word Carl said. Slides flashed on the screen behind him as he spoke about the discovery of the virus, its characteristics and how the human immune system could easily kill the virus so they had to cloak it in the right type of protein so the body would receive it. He explained the functional aspects of the virus, how it inserts itself into a damaged neuron and is able to cluster together to replicate electrical signals and DNA markers to join to adjacent cells, allowing impulses to pass through the new tangle and synapse created. Members in the audience were mesmerized by the presentation. Flashes of colored slides played on the screens as Carl spoke. There was no sound in the auditorium except for an occasional cough, the sound of the air conditioner and Carl's booming voice.

When he finished, the audience gave a uniform standing ovation. There were no questions. They were stunned. Medical science at its best. Superb. Bravo!

Carl gathered his papers and walked back to his seat. There was no time to take questions since Owen was next to cover some of the statistics of the phases, the efficacy analysis and the more mundane aspects of the development; all important, all relevant.

Owen approached the lectern and glanced to the projector operator, signaling the start of his presentation. "Thank you, Doctor Kruger. Ladies and gentlemen, my part of this presentation is to detail the specifics of the development, including the statistical information relevant to efficacy testing and advancement." The screen behind Owen flashed a timeline of the product development. "To begin, it is important to detail the timeline of the development of the product and testing with the analytical results.

"On November 19th, 2014 I received a call from a fellow neurologist who studied at Johns Hopkins from 1993 to 2004; Doctor Adam Daniels. He contacted me regarding three patients that he and his fellow neurologists examined with remarkable memory improvements." Owen continued to detail the dates of his first meeting with Carl to examine the 112 scans; January 14th, 2015, the first sample extractions of the patients; February 4th thru the 6th, 2015, the discovery of the virus by Imar Spaan, February 19th, the discovery of immune deficiency, February 20th. Owen had no notes and never looked at the slides.

He continued through his litany of dates and statistics to Leroy the rat, April 24th, 2015. Leroy mastered two hundred and seven decisions in eighty-four seconds with no errors. June 12th, Albert, the monkey, one hundred picture pairs, fifty-two seconds.

Carl watched as Owen continued to rattle off numbers, dates and statistics without the use of notes. When Owen got to the percentages of the testing samples, success rates, number and percentages of patients who died from other causes and the names of each of the other causes, Carl knew something was terribly wrong.

Owen had to have been inoculated. No one could recite those numbers and statistics in the order of the slides without error and without notes. No one. Except Quinton, Bonnie or Mac.

But Owen did.

When Owen finished, the audience gave a standing ovation as well, not for the content of the material, but for the delivery of the presentation. Owen was oblivious and thought they were applauding the material as he walked to his seat.

"That was pretty impressive, Owen," Carl said.

"Thank you, but I think it's kind of boring myself."

Owen leaned back in his chair and smiled as the Chairman approached the lectern and thanked the two doctors for their unique and informative presentations.

Carl noticed that Owen never straightened his hair; not once.

Inklings

November 21

THE PLANE WAS full of people flying across the Atlantic Ocean to Europe. Most of the people in first class were businessmen and executives. Carl, Alice and Joey were somewhat out of place amongst the first class passengers. Joey was nine years old, going on ten. He had a plethora of video games, movies, books and electronic activities on his PED[21]. His teachers digitized his homework so he wouldn't fall behind during his trip to Sweden.

Alice was reading another book on her pad. She just finished one on the first leg of the flight and was delving into her second when Carl interrupted her. "Looks like you have taken on quite an interest in reading."

"I have. It's a way to keep my mind occupied and busy, otherwise I tend to get bored just watching stuff. I guess I need to activate my imagination a little more than normal."

"That's good. Reading is a super hobby. I get tired of reading medical reports and papers."

"Maybe you should take some time off on this trip and just relax? You know you deserve it."

---

21 PED—Personal Electronic Device—used for phone, Internet, photography, voice, video conferencing, data storage, and other electronic applications.

Carl closed his notepad. "I don't know that I can. I've been scheduled to give my lecture on Alzheimer's and Cerebtol two days before the banquet and ceremony. I have to be ready."

"You are ready, Carl. You've been living and breathing this for three years."

"Possibly so, but this is so big. It's overwhelming." Carl opened his notepad again.

Alice reached over and closed his pad. "I think you should wait until you meet Imar and discuss who will present what. No need to do his work for him, too. You need to share the love," she said as she smiled.

"Share the love. Huh. Feels more like share the pain."

"I think they'll just hang a medal around your neck, hand you a check and send you off on your merry way," Alice said with a smile. "Ask Imar. He knows all about this, having won it once before."

"Did you know he is only the fifth person to be a dual laureate?"

"Yes, I did," Alice replied. "It's all over the news. And this will be your first."

"Yeah. Hard to believe." Carl leaned back in his chair watching Joey press the screen of his pad. "This will be an educational adventure for him."

"And for us," Alice replied. "When's the last time we had a month off and most of it paid for by the University?"

"Never." Carl took a deep breath and closed his eyes. Alice returned to her pad and book while Carl continued. "I guess I can wait until I meet with Imar." Carl listened to the hum of the jet engines and soon dozed off to sleep.

—*—

The Alzheimer's Disease Research Center at Johns Hopkins was very quiet. Most people were either gone or ready to leave for the Thanksgiving holiday weekend. Many tacked a few weeks of vacation onto the week and took nearly a month off. There were no major projects underway. Cerebtol was out, distributed and well established. Periodic reports funneled into the department through Carl's office, but with him and his secretary out, as well as most of his research team, the reports, emails and all phone calls were forwarded to Owen for review and processing.

Owen was scanning his emails when one caught his attention.

Date : November 21, 2018
From: Jennifer Coogle
To: Carl Kruger
Subject : Dad's condition

Doctor Kruger;

I think something is happening to my dad. At our Phase One introductory meeting, you asked that if we saw any behaviors or conditions out of the ordinary with the recipients of the drug, to contact you with specifics. I am doing just that. My dad has been experiencing some minor memory loss for about a week. It's probably nothing to be concerned about, but it seems like it just appeared out of nowhere.

Owen could feel his stomach drop. The palms of his hands started to sweat, so he wiped them on his shirt. He straightened his hair and continued to read.

This morning dad was unable to put his shoes on properly. He had the wrong shoe on the wrong foot. He complained about his feet hurting and I discovered what he did. We laughed about it, but then I watched him open the refrigerator and stare inside, then he closed it and got the phone off the counter. I asked him what he was doing and he just said 'nothing.' He's been misplacing his wallet and small items like that. I'm worried something is going on. What should I do?

Jenny

Owen pushed himself away from the desk and walked to the window, then returned back to the desk and stared at the message. He walked back to the window. "Damn!" He ran his hand over his bald head, messing up his hair. He didn't care. "It's got to be something simple. Something explainable," he said to no one. He walked back to his desk.

Owen stared at the screen and the words "memory loss." They appeared huge to him. Owen closed the email and left his office. He walked down to Carl's office and to his desk. Carl's receptionist was on vacation as well. There was no one around to see Owen step into Carl's office and bring up his computer. Owen opened the email again, took a deep breath and started typing.

Jenny:

I don't think it's anything to worry about. As I recall, your father is aged, and the memory changes could be from diet or vascular changes. I think you should observe him over the next month, document the activity, and if it continues, send the information to me right away. I'll schedule an examination with the neurologist that referred him to the program if it continues. Until then, have a Happy Thanksgiving.

Doctor Kruger

Owen pressed the "send" button and the email was on its way. He deleted the incoming email file and the sent email file. Then he turned the computer off and returned to his office. He sat at his desk and deleted the copy of the incoming email from Jenny. He opened his investment account at the online trading company and reviewed his portfolio. He had 400,000 shares of Antrole pharmaceuticals at forty dollars a share; a whopping $16 million from a $100,000 investment, courtesy of Alicia Gold. He placed an order to sell 10,000 shares at market. The order was filled in seconds, and $220,000 was deposited into his holding account waiting for the settling of the sales.

Owen straightened his hair as he closed the program, shut down his computer, grabbed his briefcase and notes, and headed out the door.

November 28

THE PASSENGER IN the taxi admired the buildings decorated for the holidays with lights in barren trees along walkways, lights along the window frames of ornate buildings, and colorfully decorated evergreen trees scattered throughout the city. The taxi drove along the waterfront, turned down a couple of streets and pulled up in front of a five-story white building with tall, glass windows framed with teal awnings and sculpted cornice along the top edge. The Grand Hotel of Stockholm was masterfully decorated for the holidays and ready for the Nobel Prize ceremonies. The hotel was the host for the Nobel recipients and the centerpiece of the city located next to the royal palace.

Imar Spaan had been there once before in 1996 when he won the Nobel Prize jointly with Peter Doherty and Rolf Zinkernagel for their discoveries concerning the specificity of the cell mediated immune defense. The hotel looked exactly the same. The city looked the same. The only things that changed in the past twenty years were the cars and the wrinkles on his face. "Thank you, sir," Imar said as he handed the driver some money.

A bellman was holding his car door open and had an umbrella waiting to receive the doctor. "Welcome to the Grand Hotel, Doctor Spaan."

Imar was quite impressed that they knew his name as he exited the cab. "Thank you, sir," he said as he held his hat and briefcase, protecting them from the blustery wind. "Chilly night, isn't it?" he said.

"Yes, Doctor Spaan. It is. We are expecting snow flurries and some sleet this evening and showers tomorrow." The bellman closed the door and escorted Imar to the front door where the doorman opened the door. "Doctor Spaan," he said. "Right this way, sir."

The doorman escorted Imar into the lobby of the hotel. The royal blue carpet with gold inlay was inviting and soft. The crystal chandeliers sparkled and cast their light on the carpet and plaster walls framed with rich, dark walnut. The columns framing the entry to the main lobby were capped with gold inlay carvings. A small, silver table stood directly under the chandelier in the center of the room. Three tall vases, each progressing in height, sat on the table with purple and white dahlias on long stems. The beauty of the room was captivating. Imar smiled as he surveyed the splendor.

"Right this way, Doctor Spaan." The doorman escorted Imar to the receptionist. She was a young woman standing behind the walnut counter. "Doctor Spaan, I presume?" she said.

"Yes. I am. I am also quite impressed with how everyone knows who I am," Imar said as he sat his briefcase on the floor.

"Doctor Spaan, we seldom have a dual recipient of the prize, sir. You are very much an honored guest," she said with a smile.

Imar blushed slightly. "I did not expect this."

"Sir, we are honored to have you here," she said, "and are committed to making this the most memorable stay of your life."

Imar smiled as he looked around and saw two bellmen taking his bags to his room and another bellman standing by to escort him upstairs. "I believe you are, my dear."

The receptionist handed Imar the key and a bellman escorted him upstairs to his room. When they walked in, Imar stopped. "Are you certain this is *my* room?"

"Yes, sir. Please enjoy your stay. And if you need anything, anything at all, please just lift the handset. It will dial your personal manager who will meet your every need." The bellman turned and left the room before Imar could hand him a few dollars. Imar looked at the bills in his hand and stuffed them back in his pocket.

The room was magnificent. It was on the seventh floor of the hotel addition facing the waterfront. A large balcony provided ample views of the boats and ferries navigating the waterway. The two bedroom suite was fully equipped with a kitchen area, stocked bar, beds, tables, plush chairs, sofas, holographic and conventional televisions and a complete business center.

Imar walked to the window and gazed out to the waterfront. He turned back and stared at the enormous expanse of his room. "I should have brought some relatives," he said to himself. He laughed and began to unpack his two measly bags into the nine drawer dresser in the master bedroom. He opened one drawer, dumped the bag of underclothes in, and closed it. The other eight drawers remained empty.

Imar took the second bag to the closet and started to hang his clothes, realizing many were wrinkled. He picked up the phone and called his personal manager for room service to press his clothes. He handed the three shirts and two pants to the lady who bowed slightly and took the items. When she left, Imar sat in the overstuffed chair facing the balcony and fell asleep.

—*—

Owen walked into Carl Kruger's office and turned on the computer. He decided it was best if he just started checking his computer first every day to intercept the emails and voice mail messages. Owen looked at the screen. Another email from a different patient. That made three this week. Owen responded to the email as 'Carl' and deleted all trace of it. Next, he checked the voice mail messages; none. If there was, he wasn't sure how he would handle that. He'd have to call back and make some promise to buy him more time. That's all he needed right now: time.

And he had about thirty days left until Carl and his team returned from their long, overdue vacation.

—*—

"Just a moment," Imar said as he woke from his short nap and slowly moved to open the door to his room. "Carl! Alice!"

"Imar!" they both replied.

"Uncle Imar," Joey said as he ran to the old man.

Imar bent over and embraced him. "Joey. This is such a pleasant surprise to see you here," he said.

"We are so glad you made it, Imar," Alice said. She stepped into the room and gave him a big hug and kiss on the cheek.

"Thank you, Alice. It's so nice to see all of you," Imar said as he shook hands with Carl. "Especially you. I was afraid I would have to do the entire speech on my own."

"Imar. You could do it easily. You don't need me," Carl said.

"On the contrary, my dear man. I most certainly do." Imar turned and waved the family in. "Come in. Come in to my humble abode."

"Doesn't look too humble to me, Imar," Alice said.

"They must pay you well on retirement," Carl chimed in.

"No, this is courtesy of the Nobel committee. Seems like a second time laureate is entitled to some perks," Imar said as he smiled.

"I would say so," Carl agreed.

The four of them chatted about family, travels, work, and migrated to the upcoming speech and prize presentation. Carl and Imar agreed they would do everything together, from the speech to the prize. Imar said he would speak to the program directors and ensure their ability to do so since it would seem he had some pull.

After an hour of chatting and planning, the group agreed to head off to dinner in the hotel lobby. After all, it was one of the best restaurants in Stockholm, was paid for by the Nobel committee and was close. The latter was the best reason from Imar's point of view. He was tired after the flight. His age was catching up to him, or so he had determined.

It was a fabulous meal and a wonderful evening. It was a grand preview to the upcoming events and ceremonies.

Euphoria and Despair

December 8

THE STOCKHOLM CONCERT Hall was packed with doctors, scientists, dignitaries and media to hear Carl Kruger and Imar Spaan speak about the discovery and development of their new medicine: Cerebtol. The chairs on the stage of the hall were arranged for the big night of the Nobel Prize Award ceremony. Ten chairs were in a semi-circle stage left. To the right sat five blue, high-backed chairs with gold trim, resembling a throne design for the King of Sweden and the royal family and hosts. Behind the front rows three additional rows of chairs were arranged in semi-circles on platforms for the Nobel laureates and nominees. A large gallery for the orchestra was above the stage on the next tier and above that hung the blue and gold flag of Sweden. In the middle was a copper medallion with the bust of Alfred Nobel, the inventor of dynamite, and the current year; 2018.

Tonight, however, Carl Kruger and Imar Spaan were the only occupants of the ten chairs stage left. Only the master of ceremonies accompanied them onstage. The orchestral pit and throne chairs were empty.

The audience sat in the sweeping auditorium. The scene reminded Carl of his presentation to the Neurological Society with Owen Pitke. This time he was with his friend, Imar Spaan, and excited to have him as a part of history.

A lectern stood center stage with a bronze bust of Alfred Nobel directly behind it. The Master of Ceremony stepped to the lectern and introduced Carl much like the Chairman of the Society did in July. The list of credentials, accolades and accomplishments for Carl and Imar were impressive. Carl was the first speaker and approached the lectern to a thunderous applause.

"Thank you, ladies and gentlemen." Carl scanned the audience. "My topic is Alzheimer's Disease and the development of Cerebtol; the newest medicine proven to effectively reverse the disease and brain cell damage through surrogate neurogenesis." The screens to the sides of the stage were lighted as Carl began his presentation of the development of the medicine and, more specifically, the characteristics of a new virus that mimicked a live cell through replicated electrical impulses. The audience was glued to his every word. The presentation was flawless.

"The virus was easily destroyed by the body's normal immune defenses. Therefore, thanks to the brilliance of Doctor Imar Spaan, we were able to develop a cloak for the virus. Doctor Spaan?"

Carl stepped back from the lectern as Imar approached. The crowd applauded heartily as the Nobel winner of 1996 approached the microphone. "Thank you," he said in his Austrian accent. "It has been many years since I stood on this stage and I never imagined I would ever be here again. I guess the pool of applicants must be shrinking." The audience laughed as Imar adjusted his notes. "As Doctor Kruger so noted, the JC-2 virus was easily destroyed by the body's immune system. Our attempts to fool the immune system into accepting the invader were numerous and futile. We realized the virus was using the acinetobacter bacterium as a host transporter into the body. Our challenge was to utilize the host until it entered the body and cloak the virus inside the host so that when the host ruptured, the virus would be absorbed into the gray matter in the brain without impediments. We used a combination of enzymes and proteins that mirrored the enzyme and protein structures of…"

Imar continued to explain the intricacies of the cloak shell that he created for the virus. Carl could see the audience was captivated with Imar's theory, his technique, the conclusion and solution that he mastered. It was the first time that Carl saw Imar for what he truly was; a humble genius who had changed the world.

When Imar finished, the audience stood and applauded. Carl and Imar stood center stage soaking in the accolades from their peers.

It was a dress rehearsal for the big event in two days: receiving the Nobel Prize in medicine.

—*—

It was more than two weeks ago when Owen received his first notification that something might be going wrong with the Cerebtol Phase One test group. Over that time, Twenty-two phone calls and seventeen emails came to Carl and Owen. Owen was meticulous to intercept all emails and calls and respond to each as though it was Carl Kruger. Each response was, *"Let's give it some time, maybe thirty days. If it continues, please let me know and we will arrange some follow up tests."* That would give Owen plenty of time to liquidate his stocks and major assets, plan his trip to a non-extradition country and run.

All he was waiting on was the Swiss account.

Owen stopped by the post office on the way home to pick up his mail. He pulled the mail out of the post office box and saw the envelope from Micheloud & Co., a Switzerland based company that offers Swiss brokerage services, which include account openings. Owen quickly ripped the envelope open. Inside was a letter from the brokerage company addressed to him.

> *Mr. Pitke;*
>
> *It is with great pleasure that we send this notice of confirmation to you. The Bank of Switzerland has accepted your application to open a premium checking and savings account with their institution. A bank representative will contact you shortly by mail to sign the final documents. Please follow their instructions closely and completely, and your account should be opened within two days.*

Owen read the rest of the letter and was thrilled to see he would finally have a place to transfer his funds from the sale of stock in Antrole Pharmaceuticals.

It was time to go home and pack.

December 10

IMAR SPAAN WAS sitting in the hotel lobby in a French Provincial chair with dark wood trim, waiting for his friends. The elevator bell rang, drawing his attention. Alice Kruger stepped out wearing a long, black evening dress with black silk trim across the bottom. She wore patent leather, open toed shoes that sparkled. She had a handbag draped across her chest that blended into the color of her dress, making the handbag hardly noticeable. Her hair was beautiful, her makeup flawless. Her shawl covered her crutch to where it was hardly noticeable. She was holding onto Carl's arm and barely used her crutch. Joey was wearing a black suit and tie that scratched his neck.

Carl was dressed in a black tuxedo that fit his muscular body perfectly. His white shirt and bow tie contrasted sharply to his dark, brown skin. His gold cuff links and gold and diamond ring sparkled. His head was perfectly shaved, and his Van Dyke beard perfectly trimmed.

Imar stood as the trio approached. "It appears the king and queen have just entered the lobby," he said with a smile. "And prince," he said as he leaned down and gave Joey a hug. "Alice, you look beautiful, as you always do."

"Thank you, Imar. You do too." Alice leaned forward and kissed him on the cheek. His hair was slightly combed, but still had a wild air about it. His tuxedo was nice, but seemed a bit wrinkled. His shoes were already scuffed. "I didn't know you would clean up so nicely," she said as she straightened his tie.

"Neither did I," Imar quipped. "Shall we?" he said as he motioned to the door.

The four of them were escorted by their limousine driver and chauffeured to the Stockholm Concert Hall where the Nobel Prize ceremonies would soon take place.

—*—

It was early afternoon in Baltimore. The snow was falling as the cold wind whipped through the trees. Owen stood by his window looking out, wondering what it was like in Thimphu, Bhutan right now. He picked up his PED from the desk and touched the screen a few times. The weather forecast for Thimphu popped up on the screen. "Cold, but dry," he mumbled. He tapped the screen a few more times, checking his account balances in Switzerland and other offshore accounts. "Good." He tapped the screen again, checking his flights.

Owen put the pad down and looked around the room. "In a couple of days this will all be behind me."

The emails and phone calls from several Phase One recipients seemed to slow down some. Maybe his request to give it some time worked. That's all they needed, just like him: time. Owen needed time to get his stocks liquidated, which he did. Just over twelve million dollars was sold from Antrole pharmaceuticals alone. Then he had to liquidate his physical holdings in gold and other investments. It took time, but he did it. All told, just more than fifteen million dollars sat in his newly created Swiss and offshore accounts. The wire transfers went flawlessly.

Next was his trip to Bhutan. Owen chuckled as he thought most people didn't know the country even existed. Being one of only four countries with no diplomatic relations with the United States, thus no extradition relations, it was the perfect place for him to get lost over the next few years. The country is primarily Buddhist, is poor in relation to our wealth and is slow with technological developments. Just more than a hundred thousand people live in Thimphu, the largest city of Bhutan and he would, by far, be one of the richest. He already made contact with a diplomat in New Delhi who would help him integrate into the little country.

"Maybe I'll buy a little temple and hide somewhere in the mountains," he mumbled.

—*—

Alice sat in the front row of the Stockholm Concert Hall auditorium. Joey sat next to her. They could easily see that the chairs arranged behind the royal family were empty. She wondered where Carl and Imar were. The orchestral pit was full of world class musicians playing Sweden's national anthem. Everyone rose and stood quietly as the music played and the royal family entered to stand in front of their blue, high-backed chairs with gold trim. The audience applauded respectfully and took their seats along with the royal family when the orchestra finished.

King Carl Gustav XVI of Sweden was surrounded by the Queen, his two princesses and Prince Carl Phillip of Sweden. The King wore a black tuxedo with large medals strung down the left side of his jacket. The Prince wore a similar tuxedo but had one medal to display on his chest. The queen and the princesses were dressed in long, flowing silk dresses. One wore lavender, another white and the Queen in brown with gold lace trim. Each lady wore a crown and had a banner draped across the front of their dress from shoulder to hip. The queen wore long white gloves and a necklace of emeralds and sapphires.

The audience was packed with dignitaries, special guests and fellow scientists of every nature. The President of the United States sat in the front row a couple of seats away from Alice. His secret service agents were scattered along the walls and doorways. Security was tight, but hardly noticeable.

The orchestra played a Beethoven melody as the audience stood again and watched the Nobel Laureates enter center stage. Two lines of impeccably dressed scientists and doctors from every field entered from both sides backstage, down the center aisle, splitting to the right and left of the stage and into their seats arranged behind the royal family on both sides of the stage. Carl and Imar led the procession and took it all in. They were seated in the first two chairs stage left across from the royal family. They looked magnificent dressed in their tuxedos.

Alice was so proud of her two men.

The Master of Ceremonies took to the lectern and greeted the audience, laureates and royal family. He spoke of world peace, global accomplishments and the need to continue to strive for excellence in everything we do. His fifteen minute speech seemed to drone on. The orchestra played another melody, this one by Swedish composer Joseph Kraus.

Each category of recipients had a dignitary in the field speak of the laureates' accomplishments, discovery and significance to the world of science. Physics and chemistry were the first categories to be recognized. The speaker ended their presentation with their gratefulness for their accomplishment and their honor to present them with the Nobel Prize in their field. The laureate would stand, along with the rest of the assembly. A person seated behind the royal family would hand the King the awards. The laureate would walk to the center of the stage where a gold "N" was woven into the carpet. The King would hand them their prize with the left hand, shake the right hand and bow slightly. The orchestra would play a ditty as each recipient would receive the prize, bow to the King, to the other laureates and then to the general audience in a three point circle. The recipient would take their seat to the applause of the assembly.

After physics and chemistry, the orchestra played Puccini from Madame Butterfly accompanied by a soloist, Greta Sahr, who was standing in the center of the top balcony under the flag of Sweden. When she finished, the audience politely applauded.

A short, middle-aged man with graying hair and a full beard stepped to the lectern. He was Professor Hans Kroleg, Secretary of the Nobel Assembly of the Karolinska Institute of Sweden. He was the presenter of the Nobel Prize in Physiology or Medicine.

Professor Kroleg adjusted his glasses as he began to read from his script. "Tonight will be a memorable night for many of you. This will be an evening you will not soon forget. Particularly you, the laureates of the Nobel Prize," he said as he motioned to the people seated to his right. "You may very well remember this moment the rest of your lives. Unless … unless you suffer from a debilitating malady called Alzheimer's Disease.

"Millions of people throughout the world suffer from this scourge. The disease robs people of their precious memories without regard for age, sex, skin color, religious preference, nationality or social status. Paupers and princesses, men and women have fallen victim to its power to destroy.

Alzheimer's Disease attacks the very life of brain cells and turns them into useless tissues incapable of transmitting neural activity through the synapses of the brain, thus destroying our ability to store or retrieve memory. The disease progresses without hindrance, leaving victims unable to care for themselves, even the most basic functions of life. It robs them of their precious memories and their dignity with impunity.

"The disease has been studied voraciously for decades with some progress. However, it was not until nearly four years ago that a journey began to uncover a remarkable virus that brought life into the dead and dying neurons through surrogate neurogenesis. The virus, known as JC-2, was discovered by Doctor Imar Spaan and Doctor Carl Kruger of the Alzheimer's Disease Research Center of Johns Hopkins University School of Medicine in Baltimore, Maryland." Professor Kroleg explained the intricacies of the virus and how it impacts the damaged brain cells. He elaborated on Doctor Imar Spaan's significant contribution to the research of developing a protein and enzyme cloak for the virus to defend the attacks from antibodies and migrate to the brain where it took hold and multiplied in the damaged cells. The audience listened to how history was made.

Carl and Imar sat stoically listening to the tale of their adventure. Carl occasionally glanced at Alice and Joey seated in the front row and smiled.

As Professor Kroleg finished the explanation of the discovery and development of Cerebtol and the functions of the virus, he set his notes aside and turned to the laureates. "Carl Kruger. Imar Spaan." Both men stood as he called their names. Carl towered over Imar. They were an odd couple.

"Your research has identified a cure for Alzheimer's Disease. Not only has your discovery provided a cure for this dreaded disease, but your discovery of viral cell infusion has offered hope for mankind in its quest to cure paralysis, dementia, Parkinson's Disease and other neural degenerative diseases and injuries. On behalf of the Assembly of the Karolinska Institute, I want to convey our warmest congratulations. Doctor Imar Spaan, Doctor Carl Kruger. May I ask you to please step forward to receive the Nobel Prize in Physiology or Medicine from the hands of his majesty, the King."

A man seated behind the royal family stepped forward and handed the King a plaque with a red box on top. The King took the items and stepped forward to center stage where the gold "N" was woven into the carpet. The entire assembly stood.

Carl slowly walked to the "N" first and met the King. The King handed Carl the plaque with the box on top with his left hand and reached over the top with his right hand to shake Carl's hand. The orchestra played a ditty as Carl accepted the prize. The King bowed slightly, as did Carl in return. Carl turned to the other laureates and bowed slightly, then turned to the audience and bowed. He could see Alice and Joey in the front row clapping furiously. He was elated. Alice and Joey were so proud.

Carl returned to his seat as Imar started to walk to center stage. "Knock 'em dead," Carl whispered as Imar passed.

"What?" Imar whispered, not understanding what Carl meant. He looked confused for a second as he brushed off the remark and met the King center stage, received the prize, shook hands and bowed to the King, laureates and assembly. He then returned to his seat and stood next to Carl. "Why would I hit him?" Imar asked.

Carl laughed.

Everyone took their seats as the orchestra played another tune, several more laureates accepted their prizes and Carl and Imar watched the show from the stage. When all of the laureates had received their prizes, the assembly stood as the orchestra played Sweden's national anthem. When finished, the royal family exited the stage and the laureates milled about, shaking hands, hugging and congratulating each other.

Within a few minutes people started filing out to their cars to be transported to the Stockholm City Hall for the Nobel Banquet.

—*—

A long line of cars framed the entrance to the Stockholm City Hall. The large brick building with a twenty-five story watchtower in the corner was tastefully lighted to highlight the arched entryways and corridors around the building. The Blue Hall was decorated with long tables set for a king. Sparkling dinnerware of the finest china with green borders and gold trim, silver flatware, crystal goblets with gold stems and the finest linens greeted each guest with a personalized name card placed on top of the water goblet. Alice loved the decorations. "Oh, Carl. Look at this table!"

"Yes, it is quite beautiful," Carl said as he held the chair for Alice. Joey sat next to his mother and Imar next to Carl. "I never thought we would see this, Imar."

"Though this is my second time here, I must say it is still very humbling." Imar looked around the room. "The President of the United States is over there," he said as he pointed.

"And I see the King of England there," Alice said as she nodded to her right.

Joey started to reach across the table, and Alice stopped him. "Joey, this is a very special night, so we have to use manners that we are not accustomed to."

"Huh?" he said.

"Don't reach for anything, son. Everything will be passed to you and your mom will help you with dinner, O.K.?" Carl said.

Joey knew this was no ordinary dinner. "Yes, dad."

"Well, Imar. You just received almost a million dollars as your prize award. What are you going to do with all of that money?" Alice asked.

Carl answered before Imar could speak. "I know we are going to pay off some debts, set up a college fund for Joey, put some in retirement and then maybe go on a long vacation," he said. "Like a year's sabbatical."

"Sounds good to me," Alice agreed.

"Things are a little different with me, Carl. Since I'm not married, I just enjoy working and a few hobbies."

"Imar, I would think there would be someone eager to meet someone like you," Alice said.

"Oh, there was, a long time ago."

"There was? You never mentioned that," Carl said.

"Didn't really see a need to, until now, I guess," Imar replied.

"You don't have to talk about it, Imar if you don't ..." Alice said as Imar interrupted.

"Oh, no. It's fine." Imar looked at the three and continued. "You are my friends and almost feel like family. We have spent much time working together over the past few years. You remind me a lot of my wife, Alice. Caring, understanding, patient," he said.

"Patient? Are you sure you mean Alice, Imar?" Carl said as he laughed.

"Yes. My wife was wonderful. Anna. And my daughter Carli was such a bundle of joy."

"What happened, Uncle Imar?" Joey asked.

"Joey! It's not polite," Alice said as Imar stopped her. "No. That's fine," he said. "It's O.K., Joey. They were both killed by a very angry man."

The table fell silent. "I'm, sorry to hear that, Imar," Carl said.

"Me, too, Uncle Imar."

Alice was stunned.

"That was a long time ago and now I have very fond memories of them and look forward to that day when I will see them again, in heaven." Imar smiled as he placed the napkin on his lap. "So, since I don't have quite the same responsibilities as you, I plan to give my award to the SOS Children's Villages."

"SOS?" Alice asked.

"Yes. It's a charity founded in Austria in 1949. It's the world's largest organization that takes care of orphaned and abandoned children and gives them a home, a family and an education. I have been a member since 1998; the year my family was taken."

Alice nearly came to tears as she saw the compassion that Imar had for lost children. "That is just wonderful, Imar."

"I knew you were a good man, Imar. That is why I have always liked you," Carl said as he slapped Imar on the shoulder.

The servers brought several bottles of wine and some sparkling cider for Joey. After the glasses were filled, Carl raised his glass. "A toast." The four of them lifted their glasses. "To Doctor Imar Spaan, the Nobel Foundation and the future."

The four clinked their glasses together.

This was certainly a night to remember.

December 17

THE KRUGER FAMILY was walking quickly along the people mover conveyor belt to gate thirty-four at terminal one in the Paris airport. Their flight had a minor change because of mechanical troubles. "Keep up, Joey," Alice said as Carl led the procession along the belt. "Our flight is boarding and we don't have much time."

"I am, mom. I'm trying." Even on crutches, Alice was staying ahead of the tired, little boy dragging his small bag.

Their flight from Sweden went as planned, but the layover was much longer than expected. They had a choice of either taking a later flight with two additional connections, or trying to get this one that went directly to Newark, New Jersey. They chose the direct flight, though it didn't leave them much time to board.

"This way," Carl said as he walked off the end of the belt and rounded the corner towards the gate. He stopped to look back and make sure Alice and Joey were catching up when something caught his eye.

"Owen?" Carl whispered.

Owen Pitke was boarding a flight to India across the terminal from Carl at gate thirty. He was next in line to hand his ticket to the agent when he looked back and saw Carl standing near the corner of gate thirty-four and the conveyor belt about fifty yards away. "Next." The agent's voice broke his gaze as he handed her the ticket. He glanced back and saw Carl moving a little closer for a better look.

"Thank you," she said as she handed his ticket back to him. Owen turned, smiled, straightened his hair and saluted Carl with two fingers. He turned and walked down the ramp to board the plane.

"Owen," Carl yelled, but he didn't stop.

"What?" Alice asked.

"Did you see that?" Carl said. "It was Owen, Owen Pitke."

"You're imagining it."

"No, it was. He even waved," Carl said as he walked back to Alice and Joey, who was out of breath.

"Final call for flight three thirty-two for Newark, New Jersey at gate thirty-four," came the voice over the speaker.

"Carl, we have to go," Alice said.

Carl looked back at the gate, but only a few more people were in line. Owen was nowhere to be found. Carl noticed the sign at the gate said, 'Delhi, India.' "India?"

"Carl, now," Alice said as she hobbled toward the gate, leaving Joey and Carl to catch up.

—*—

The hum of the jet engines was soothing. Joey was content playing games and listening to music on his PED. Alice was reading a book on hers. Carl forgot to charge his pad before they left, so he was resigned to scanning the local magazines and newspaper provided in first class. At least they were current periodicals and he appreciated that. There was nothing worse to him than reading old, out-dated magazines left at a doctor's office reception area or an airplane. He often wondered how many sick people touched those and what germs or viruses were being passed along to other unsuspecting people.

As he flipped through the magazine, Carl came across an article about the Nobel Prize ceremony in Sweden. "Hey, look at this," he said as he pushed the magazine over to Alice. He pointed to the picture. "That's me."

Alice lowered her pad. "Yes, nice shot," she said as she smiled and went back to her pad, uninterested.

Carl read through the article about the different categories and recipients, their accomplishments and stories. One was a scientist whose parents were children interned in the German concentration camps during World War Two. They watched thousands of Jews murdered, but they survived, were released by the allies and soon married. He was born to eventually discover the interaction between baryonic dark matter and a new super-symmetric particle. It was way over Carl's head, but what caught his attention was that someone could be that brilliant.

Then he came to the prize in Physiology or Medicine. The article was about him and Imar, their discovery and the millions of people inoculated for Alzheimer's Disease. The article compared them to Louis Pasteur and the discovery of germs, James Watson and Frances Crick and the discovery of DNA structures, and Jonas Salk with polio. Carl laid the magazine on his lap and closed his eyes. It was the first time he truly understood the significance of their discovery. They were now a part of human history.

December 18

IT WAS ALMOST noon. Jet lag from the trip to Sweden was weighing heavily on the Kruger family. Joey's school was on winter break. Alice had no plans for the holidays; just having Christmas at home and maybe visiting a few friends and, of course, her mom. Carl wasn't scheduled to return to the office until after Christmas, but seeing Owen at the airport was troubling.

Carl called Owen's cell phone. There was no answer. The phone went to voice mail and the mailbox was full. Carl found that odd, but if Owen was in a country with no reception, that could be possible. *But why go to India? Owen never talked about traveling there.*

Next, Carl called the office to see what was going on. The receptionist told him Owen was on vacation and that she was so proud of Carl for receiving the Nobel. She watched it on the Internet and was so excited. Another fan of Carl's. He asked if there were any messages. The receptionist said Owen closed the department for the Christmas holidays and let everyone take their vacations during the two weeks. He said there was no reason for everyone to be there at the same time.

That seemed odd to Carl since Johns Hopkins had never done that before, close an entire department. Not that he was aware of. But, the receptionist was right. There was nothing going on, no pressing issues that he knew of.

Carl decided he would just spend the holidays with his family and go in next week as planned. Another couple of days off wouldn't make a difference. They gave him the time off, so he should take advantage of it. No big deal.

Or, so he thought.

—*—

It was late in the evening when Owen Pitke landed at the Paro airport in Bhutan. He couldn't see much of the surrounding area, only that it wasn't very populated. The lights only extended for a couple of miles in any direction. He picked up his bags at the small baggage claim and headed toward the exit. The bitter cold air slapped him in the face as he stepped outside. A cab sat at the curb. "English?" Owen asked. The man shook his head 'no.' Owen pulled out a page he printed off the Internet. "Thimphu?" The man shook his head 'yes.' "Taj Tashi?" Again, the man shook his head 'yes' and opened the door for Owen to get in.

Owen climbed in as the driver put his bags in the trunk.

The cab sped off into the dark.

December 21

IT WAS THE Friday before Christmas. Joey ran downstairs to see what his morning clue was for his little present. He opened the piece of paper.

15-22-5-14

He looked at the numbers over and over. They made no sense to him. He whispered them out loud, turned the paper upside down, but couldn't figure them out.

"What's your clue today, honey?" Alice asked.

"Numbers. Just numbers, nothing else," the nine year-old boy said dejectedly.

"Hmm. Maybe it's code," Alice said.

"Yeah. Code." Joey ran around the counter to get a pen out of the drawer. He counted the alphabet on his fingers and wrote down the letters corresponding to the numbers on the clue.

O-V-E-N

"Oven!" He yelled as he dashed around the counter again and flung the oven door open. Inside was a small, wrapped package. He quickly opened the package to reveal his little Hot Wheels truck. "Cool!" he said as Carl walked into the kitchen. "Thanks, dad," Joey said as he ran to his dad to hug his waist.

"Wow. Now that's a greeting," Carl said as he knelt down and hugged his son.

"This is awesome," Joey said.

"You are quite welcome, young man," Carl replied.

"What are your plans, today?" Alice asked.

"I need to go to the office for a few hours."

Alice stopped preparing breakfast. "Why? You're on vacation."

"Seeing Owen at the airport, hearing that no one in the department is there, I'm just bugged by it. I just need to go and check in. It'll just take a couple of hours. I'll be back by noon."

"Promise?"

"Yes, promise."

"Good. We're going to Jan's for dinner tonight and I'm looking forward to it."

"Sounds like fun," Carl replied, still thinking about the office.

—*—

The Taj Tashi hotel in Thimphu, Bhutan stood as a monument to the opulence of man. The five story, rust colored hotel was a landmark in the small, capital city of 90,000 people. The grounds were impeccably maintained. Large patio areas with fire-pits were surrounded by benches with pillows and ground lights. A small temple was built on the grounds for the numerous Buddhist guests from around the world. Owen Pitke looked like another dignitary or European businessman dressed in slacks, long wool coat and gloves as he walked through the lobby of the hotel. His short beard and shaved head gave him a distinguished look. He grabbed an apple from the counter as he put on his hat and scarf to protect him from the cold wind coming down the mountains. He scurried out of the hotel to a waiting cab and doorman. "I tink you are going to enjoy the tour I have arranged for you, Meester Wallace," the doorman said as he held the door open for Owen to climb in.

"Can't wait," Owen said matter-of-factly. His new name fit him well. He liked the sound of it; Meester Wallace. Owen William Wallace. He wondered if the guy ever watched the movie Braveheart. *Did he know who Mel Gibson was?* Didn't matter. He had a new name and he was a very rich man in a very poor country. He was above everyone else because they were peasants, inferior beings there to serve him because he had great wealth and was brilliant.

The clean-dressed Indian man walked to the other side of the car, climbed in and sat next to Owen. He was middle-aged, thin, had black hair and big, brown eyes. His smile was infectious. He spoke English well, but clearly had an Indian accent. "O.K., driver," he said.

The street was bustling with small cars driving on the wrong side of the road. At least, to Owen Pitke, or William Wallace, it appeared to be on the wrong side. "Have you arranged the meetings, Fadid?"

"It is Faiz, Meester Wallace. Faiz. It means 'gain' in Sanskrit."

"Sorry," Owen sighed. "Faiz."

"Yes, sir. I have taken care of all tings. We will be touring several locations able to accommodate your need for housing. I have discussed this with the city leaders who are very much in favor of supporting your desired investment."

"I just want a big place with security where I can live out my life … like a king," Owen said as he smiled.

"I fully understand, Meester Wallace. We will have such a place for you in a short time."

Owen wasn't sure if he could trust Faiz or not. Owen was in a foreign land, and the man came highly recommended through a network of contacts he made after he and Sinclair plotted to dispose of Quinton Lemolo. Two years ago, Owen would have never imagined himself sitting in a cab with a scum like this guy, or agreeing to hire someone to kill another man, or becoming a multi-millionaire who got a fake passport and documents to change his name and hide in a foreign land. Now, here he is. "I want to make sure we keep this low key; under the radar," Owen said. The man looked at Owen curiously. "I want to keep my presence and plans private; quiet."

"I see, Meester Wallace. I will make sure that this is not well known that we are meeting with these people."

Owen watched the average looking buildings zip by as the car sped down the main road through the city. At each corner stood a policeman in blue uniform with a white belt and gloves directing traffic. The policeman had a unique manner of waving his hands to direct traffic, almost like a dance. "That's pretty impressive," Owen said as they passed the policeman standing on the edge of the colorful platform in the middle of the road. His breath showed as a cloud each time he blew his whistle or exhaled. Snow lined the streets and building fronts.

"Yes, Meester Wallace. We have no traffic lights in Thimphu; only these policemen to direct traffic."

"Lights would be cheaper."

"But more people would be out of work," the man said.

As the car drove south on the expressway and headed out of town, Owen could see the monasteries cut into the hillsides above the valley. A thin layer of snow blanketed the hills from the recent snowstorm.

They drove about six miles and turned off the expressway and started up a hill. The land was somewhat barren with only scrub brush and sparse evergreen trees mixed in the snow. The car continued until it came to a small dirt road to the north. The driver turned and continued for a few hundred yards down the mud and snow road and stopped. Two other cars were parked at a flat area near the edge of the hillside overlooking the valley. Four men stepped out of the cars wearing heavy coats with hoods, and boots.

"These men will discuss your plans to buy this land and build your house here, Meester Wallace," Faiz said. "They are the city manager, local architect, and engineer for Thimphu."

"Who's the fourth guy?" Owen asked.

"Security." Faiz smiled. "It is good to have security, Meester Wallace."

—*—

The parking lot in front of the large brick building at Johns Hopkins was nearly empty as Carl pulled in. "Looks like everyone is gone for the holidays," he said as he pulled into a space in the front row. He saw 'Director' painted on an empty space across from him, and thought once again about Owen at the airport.

Carl grabbed his briefcase and walked into the empty building and made his way to his office. He passed a couple of people who stopped him, congratulated him, said they saw him on TV or in the news and how proud they were of him and his award. Fans, more fans. Seemed like almost everywhere he went someone now recognized him and made a comment about his wonderful discovery. It was almost embarrassing, but recognition was wearing nicely on Carl. He was big, strong, handsome and humble.

Carl made his way to his office. There was no receptionist, no staff. Everyone was truly gone. He put his coat and briefcase on the sofa near the window and looked at the view. The snow-covered grass stretching to the river was beautiful. He went to his desk and turned the computer on. As it booted up he saw the message light blinking on his phone, so he hit the speaker button and called his voicemail to retrieve his messages.

"You have one hundred and twelve messages," the reply came from the auto attendant.

"What? A hundred and twelve?" Carl knew he would have a lot of messages, but this seemed a little excessive.

"First message," came the mechanical voice, "two forty-two PM, December 20, two thousand and eighteen."

"Mr. Pitke. I don't know what to do," came the anxious female voice from the recording. "My dad's getting worse. It seems like he is forgetting more every day." Carl felt his stomach turn as he leaned forward. "I really need to talk to you to see if we can get him tested or something. Can you call me back as soon as you get this? That's Jennifer Howell, 402-555-1899. Thank you."

Carl grabbed a notepad and wrote the name and number down as he replayed the message.

"Next message, one twelve PM, December 20, two thousand and eighteen."

"Mr. Pitke. My name is Harold Fisher," came the voice of an older man. "My wife was in the Phase One test group for Cerebtol. Sir, she is experiencing some memory loss that is very troubling."

Carl paused the message. "Oh, Lord, no," he whispered as he wiped the top of his head. He could feel the sweat forming. He pressed the button to continue. "I'd appreciate it if you could call me back right away and tell us what to do. She's scared. Her name is Madelyn. My number is 380-555-3822."

Carl turned the speaker phone off and stared at the blinking light. He didn't move. He didn't want to be there. He wanted to run, but there was nowhere to go. He paced to the window and back several times, then returned to the desk. He looked at his monitor, which automatically brought up his email account.

Six hundred and eleven messages.

He scanned down the messages and saw many were forwarded from Owen's email. Many were normal messages about operations or notices from various departments and vendors. He scanned down to one from Adam Holvarth.

Mr. Pitke. My mother was in the phase one test group for Cerebtol. Last week she couldn't remember what day it was.

The message continued, but Carl stopped reading.

He didn't know who to call or what to do. He felt like a caged animal. His eyes were wide, his hands were sweating.

Millions of people, he thought. *Millions.*

Carl slowly took a notepad from his desk and drew several columns on the pad. He noticed his hands were shaking. It was surreal. He felt like he was moving in slow motion as he reached to the phone and played back each message, listing the name, phone number, and any information or symptoms he heard.

The words were haunting.

"… forgets … memory … mother … dad … wife … increasing."

—*—

Hours passed. Yellow pages torn from the notepad lay strewn on the desk along with several paper coffee cups. Carl had worked non-stop to sort through the hundred and twelve voice mail messages and six hundred and eleven email messages. He was writing furiously when his cell phone rang.

"Carl. Where are you?" Alice asked.

"What? What time is it?"

Carl looked at the monitor as Alice answered, "Five-thirty. We have dinner at Jan's in an hour."

"I don't know if I can make it, Alice," Carl said as he continued to scan the monitor and messages.

Alice could hear the despair in Carl's voice. "What do you mean? What's wrong?"

"I … I'm not sure, yet. I'm still gathering data."

"Carl, it's Christmas vacation. We have plans. We have dinner …"

"Alice!" Carl yelled. "Listen to me. Something is wrong and I have to figure this out!" he said as he slammed his fist on the desk, knocking a cup over. He ignored it.

"What? What's wrong?"

"Cerebtol." Alice's stomach jumped when she heard the word. "It's … I don't know what's happening."

"Carl! You're not making sense," Alice shouted. "What about Cerebtol?"

"It's not working on some of the patients." Alice could feel her legs getting weak as she eased herself into a chair and listened to Carl. "They're regressing or something. I don't know!" The phone was silent. "Did you hear me?"

"I did," Alice said quietly.

"I need to find out who, when, and what the people are saying. I mean I have hundreds of messages here to sort through," Carl said anxiously as he started to panic again. "I don't know where Owen or anyone is right now. There's no one here. I don't have anyone …"

"Carl! Carl." Alice took a deep breath. "Do you need some help?"

"I do. God knows I do." Carl paused for a few seconds, and then said something he had never said before. "Alice, I'm scared."

"So am I." The two were quiet for a few seconds. Alice spoke first. "I'll be right there. I can help."

"I know you can, baby. I'll wait for you."

"It'll be alright, Carl," Alice assured him.

"I hope so," Carl said as he hung up the phone and rested his head in his hands.

Alice held the phone to her ear and whispered, "Me, too," and hung up.

—*—

It was dark and the roads were getting icy, but Alice managed to make her way to the university and park next to Carl's car. She was careful using her crutches on the icy pavement as she made her way to the front door where Carl met her. He let her into the building and they embraced and held each other for several seconds without saying a word.

"I'm glad you're here, baby" Carl said.

"Me, too." Alice sighed deeply as she rested her head on Carl's broad shoulder. She always felt safe in his arms. She leaned back and smiled at him. "Here. I brought you some dinner." She opened her messenger bag and handed him a sandwich wrapped in plastic. "Ham."

Carl took the sandwich and smiled. "You know me all too well. I haven't eaten since I left home." Carl took a big bite. "Where's Joey?"

"He's at Jan's. He's staying the night. I expected we will be here all night from the sound of your voice."

Carl swallowed and wrapped the sandwich up. "It's bad, Alice. I don't know what's happening. Owen's phone and email was forwarded to me. I had hundreds of messages from people connected to Phase One testing. They're forgetting."

Alice wanted to tell Carl she had the inoculation, but she was afraid. She thought it would be better to make sure the problem existed before she told him. Besides, it's probably nothing and some of them are getting vascular dementia or something. No sense adding more to his plate unless she was sure. Her memory was perfect. Maybe it doesn't happen to everyone, whatever 'it' is. "Let's go find out why," Alice said as she held onto Carl's arm and used a crutch to help her along to the elevator and up to Carl's office. Carl carried her other crutch and the sandwich.

December 22

THE SUN WAS breaking the horizon. Carl and Alice worked all night sorting the emails and voice mail messages into groups. They researched each person to identify the date of inoculation, sex, age, location, and diagnosis. Alice was a master at data gathering and mining. When she worked at the Centers for Disease Control, where she and Carl met, that was her job; to manage a team of analysts who gathered information from around the world and develop correlation models to determine causality, origination, or spread of different diseases. They called her the "Queen of Data Mining."

Last night, she was mining information directly related to her health. She was personally vested.

After sorting through the calls and messages, it was clear; thirty-seven of the patients were experiencing some sort of memory degradation. The time frames were all about three years from the date of inoculation.

A large whiteboard with a series of columns and rows drawn in different colors loomed behind Alice and Carl, who were sitting on the sofa looking out the window. The sun was breaking over the horizon and reflecting off the freshly fallen snow. It was a clear morning with a deep, blue sky.

"What do we do now?" Alice asked.

"I don't know," Carl said softly. "No one's here. I have no team."

"You have me."

Carl hugged her and smiled. "I do. Thank you for helping last night."

"Carl, we can't really do anything right now. We're going to have to wait until after the holidays."

"I'm not sure I can do that. I can still get more information and, at the very least, I need to let the board know something is brewing." Carl jumped back into research mode. He got up and erased the whiteboard behind them and wrote the next tasks down. "Alice, we need to call each person who left a message. Then we should call the remaining people in the Phase One group to see if anything is surfacing. We can tell them it's a 'health check'." Carl was writing on the board as he spoke. "I'll call our board chair, Margaret Giddings, and put her on notice."

"Isn't that premature?" Alice asked.

"No. I don't want anyone to think I held back or delayed information. Especially bad information. She needs to know, right away."

"What can I do?"

"Go home. Get some rest."

"And leave you here?"

"Yes. I'm going to make the calls, call Margaret and set up an emergency board meeting. From there, I can't do much else until we get direction from the board or a team back here."

"But you need help."

"Yeah." Carl thought for a moment and looked at Alice. "I need to tell Imar."

Alice and Carl became very quiet. Alice spoke first. "I'll see you at home. By tonight?"

"Yes, tonight."

Carl helped Alice to her feet and grabbed her crutches and messenger bag. Alice put the bag on, took her crutches, and kissed Carl on the cheek. "I can let myself out."

"Thank you," he said as she headed out the door.

"You'll get my bill in the morning," Alice yelled as she started down the hall.

Carl smiled and went back to his desk. It was time to make some phone calls.

—*—

The snow was blowing sideways. The large, wood-framed room was warm and cozy. A huge fire crackled in the fireplace as the black Labrador Retriever lay on the rug in front of the hearth. A white cat was lying on the dog, sound asleep. Imar was sitting in his favorite glider chair reading a book when the phone rang.

"Hello?"

"Imar, it's Carl Kruger."

Imar sat forward in his chair. "Carl! I certainly wasn't expecting a call from you. Merry Christmas, my friend."

"Imar, I wish it was a Merry Christmas. I have some bad news."

Imar put the book on the table and stood. "Please don't tell me it's Alice or Joey."

"No, no. Nothing like that. But, maybe worse. It's Cerebtol."

"Cerebtol? What do you mean?"

"Phase One recipients are experiencing a relapse."

"What?"

"Their memories are degrading." The phone was silent. "Imar? Did you hear what I said?"

Imar cleared his throat. "I did." The phone was silent again.

Carl broke the silence. "I don't know what to do."

"Would you like me to come out to help?"

Carl didn't hesitate. "Yes, Imar. I need you here. We need to find out why this is happening and try and correct it, somehow."

"I agree. I'll be on the next flight out."

"What about Christmas? Aren't you going to spend it with your family?"

Imar smiled. "Carl, I certainly hope so. If I leave now, I can make it there in time," he said with a chuckle.

Carl laughed. "Imar, I look forward to it. I'll have Alice prepare your room. Joey will be happy to see you."

"And vice versa."

—*—

It was nearly four in the afternoon, but the list was complete. Carl called every person who was in the Phase One test group except six. He left messages, but didn't expect their callback anytime soon because of the holidays. Besides, he had enough information to show there was a serious problem developing with Cerebtol.

Nearly all of the patients experienced some form of memory loss in the past two months; from minor to somewhat troubling. Three had more frequent occurrences of memory loss than the others. They happened to be the three who had Early Onset Alzheimer's. Every patient was showing some minor levels of anxiety. The memory losses were not associated to any particular event, action, or time. They happened at all times of the day, during various activities, for different reasons. They were all short-term losses and often recalled at a later time. It was difficult to assess if the frequencies were increasing since it was just beginning, but it appeared they were. Carl asked each patient to keep a note pad close by and indicate the date, time, and duration of the loss, if they could. He said he would contact them next week right after Christmas for another health check.

Margaret Giddings, the Board Chairman at Johns Hopkins, took the message admirably and maintained her composure over the phone. After a brief discussion, she agreed to see Carl on Monday and to call an emergency board meeting on Thursday, the 27th. She believed she would have enough board members to make a quorum by appearing in person or by video in case they had to take action. She directed Carl to have all of his information and evidence prepared for presentment to the board. She did not want to take much of their time away from their vacations, so he needed to be succinct.

Carl promised that he would.

He questioned how someone so educated and brilliant could miss the fact that they may very well be standing at a precipice of a disaster.

December 24

THE BALTIMORE-WASHINGTON International Airport was busy. Cars, taxis and shuttles were stopped in long lines in front of the arrival terminal. Carl worked his way through the line inching slowly toward the American Airlines terminal. "There he is!" Joey yelled as he excitedly pointed ahead. "There's Uncle Imar!"

Carl maneuvered his way to the curb and stopped a few yards short of where Imar stood. He was wearing his hat, long wool coat, and had one bag. Carl stepped out of the car. "Imar!" The two men met at the side of the curb and shook hands.

"Good to see you, Carl."

"And you. Unfortunately, I wish it was under better circumstances."

"As do I, Carl." Imar started to lift his bag, but Carl took it from him.

"Imar, I think we should go to the lab first and look at the messages and research I did. There are people suffering …"

Imar stopped Carl. "Carl. I am an old man. This is Christmas Eve; the day before we celebrate the birth of our Savior. *You* can't save the world in one night." He looked down at Joey. "Let's go celebrate with your family. Then, we shall talk, and work." Imar smiled as he leaned down and hugged Joey. "Good to see you, Joey. You have grown in just one month!"

Carl put Imar's bag in the trunk, and the three of them got into the car and headed toward the Kruger home.

December 25

THE ROOM WAS dark. The light from the house across the street shone through the blinds in the extra bedroom, across the hall and into the bedroom where Alice laid in bed staring at the ceiling. She could clearly see the light fixture in the center of the room. She had been lying there for a couple of hours, thinking. Her eyes adjusted to the night and she could see Carl's body lying next to her. The blankets over his shoulder rose and fell with every breath. Her mind was running a thousand miles an hour. *How can I tell him I had the inoculation? How long until I start experiencing memory loss? Is this going to destroy my mind? I shouldn't have done it. What's going to happen to Joey? Maybe Carl will find a cure. What if he doesn't?*

Alice could feel her stomach turn. She had to get up. She couldn't just lay in bed and drive herself crazy thinking of all of the scenarios that may or may not happen.

Carl turned. "Oh, are you awake?" he asked.

"Yeah. I just woke up."

"Mmmm. Merry Christmas, baby." Carl turned all the way around and laid his arm across Alice's neck.

"Merry Christmas," Alice replied. "Carl?"

"Yeah, baby."

"I … I want to tell you … uh …" Alice wanted to tell Carl she had the inoculation, but was scared.

"What? That you got me a special present for Christmas?"

Alice took a deep breath. "Yeah. It is special. And you'll get it when we open presents." She threw the covers off, put her robe on and hobbled onto her crutches. "I'm going to make us some coffee," she said sternly.

"O.K. I'll be right there," Carl said as Alice hobbled out of the room. "What's bugging her?" Carl whispered.

Alice walked into the kitchen and found Imar sitting at the counter drinking coffee. "Merry Christmas, Alice."

"Oh. Yes. Merry Christmas," she said. "Up early?"

"Jet lag." Imar watched as Alice poured her coffee and sat at the table. She was downcast. "You don't look much like celebrating Christmas this morning. Is something bothering you, Alice?"

"Oh, it's just, uh, no." She looked at her cup. She wanted to tell her trusted friend, but she couldn't. She couldn't tell anyone. "I guess it's just early."

"Yes, it is," Imar said as he took a long sip of coffee.

The two sat quietly as the sun peeked over the horizon.

—*—

The living room was a pile of wrapping paper, ribbons, bows and boxes. Joey sat on the floor and played with many of his new toys. He had games, cars, clothes and books to enjoy over the next year. Being an only child and adopted, he was showered with many gifts. Alice and Carl sat next to each other on the sofa drinking their coffee. Imar sat in the overstuffed chair watching the family scene. "Isn't this a joyous time?" he said.

"Yes, it is," Carl replied. Alice stared at the table. Carl nudged her. "Alice?"

"Hmm? Oh, yes, a great time."

"You seem to be pre-occupied with something. Are you O.K.?" Carl asked.

Alice looked at him. *Was this the right time? No, not with Joey here. She had to tell him, but when?* "Yeah, just a little tired I guess," she said as she snuggled closer and watched Joey play with his toys.

—*—

It was dark, cold and miserable outside. The wind was blowing down from the Himalayan Mountains into the valley dropping the temperature to twenty-five degrees. It was cold, but Owen was used to it. Maryland often had sub-freezing temperatures. The difference here was that he was bored, very bored. After just one week, Owen found little to do when it was freezing outside. There was one small theater in the city of 90,000 people and it only played outdated films, many in English, several in Indian. There were few nightclubs because of the strong Buddhist influence. No sports except soccer, no shows other than native plays or dance, and a few out-dated English TV shows.

Bhutan has fifty-three languages and in Thimphu, Dzongkha is the main dialect. It is based on the traditional Tibetan language. Fortunately, for Owen, English is the medium of instruction for the schools and education system with Dzongkha as the national language. The main entertainment that Owen had on a cold night like this was television. Owen was more fortunate than he realized. When Bhutan introduced television to their country in 1999, it was the last country in the world to do so. Now, the old 'I Love Lucy' reruns from the 1950's playing in the background in Owen's room were welcome noise. He thought he had seen them before, but he couldn't remember most of them if he did. Boy they were funny shows.

The only benefit of being in such a lousy country was that they had no diplomatic relations or extradition treaties with the United States. Regardless of what happened with Cerebtol, he would never be forced to leave. That was the only reason Owen picked Bhutan as the place to hide. Plus, he had a lot of money.

Owen was sitting at the table of his suite at the hotel flipping through the rough drawings that the architect put together for constructing his house. Others would call it a fortress. Owen met with the city planners, if that's what you call them, to discuss the purchase of the property and his plans a couple of days ago. He had signed documents to purchase twelve hundred acres a few miles outside of town bordering the Thimphu Chuu, or river. He picked a spot where his house would face the Thimphu Valley and the city lights. The city engineers said he would not have access to electricity from the city, but could plan a connection through the district government. Owen didn't care. He planned to have his own electrical generation system through a combination of windmills and hydroelectric generators off the Thimphu Chuu. He was going to be fully self-contained.

After all, he was rich; much richer than when he was in the United States. The ngultrum, the currency of Bhutan, was based on the Indian Rupee. Owen's fifteen million was worth an easy three-quarters of a billion rupees. Owen could buy almost anything he wanted.

He flipped through the drawings as the 'Lucy' shows played on in the background. Owen made a few notes here and there, adding windows, walls, vantage points, a lookout tower on the south end, gates, access control points, electric fence and other amenities he felt were necessary for his new quality of life. They were rough ideas and notes, but he was sure he would get the point across; it would be secure.

Every few minutes, 'The Lucy Show' would draw his attention, and he would forget about his plans as he laughed out loud, almost uncontrollably at times, and then return to his work.

Yes, they were funny shows. Strange that a few weeks ago they were a nuisance.

—*—

The table was brightly decorated with a green and red table cloth with Santas and Reindeer embroidered around the edges. A large, partially devoured ham sat in the center of the table with the carving knife and fork lying on the side of the platter. The mashed potatoes and gravy were almost gone. The pie was decimated. Imar was leaning back in his chair enjoying a glass of cider and cinnamon. "Alice," Imar smiled, "I believe that was the best dinner I have had since Anna cooked for me."

"Well, that is quite the compliment, I think," Alice said not really paying attention.

"It is," Imar said.

"I agree with Imar, hon. It was great." Carl took a sip of coffee.

"Thank you." Alice placed her napkin on the table and picked up her crutches. "If you'll excuse me for a few minutes." She stood and started to hobble away.

"Where are you going?" Joey asked. Carl and Imar watched as Alice ignored Joey's question and walked down the hall into the bedroom.

"Hmm. I think I should see what's bothering her, Imar." Carl stood. "I'll be right back."

"Joey," Imar said. "Why don't you show me some of the presents you got?"

"Sure, Uncle Imar!" Joey dashed from the table into the living room and started pulling his presents from under the tree to show Imar, who followed slowly behind him.

Carl walked into the bedroom and closed the door. Alice was across the room in the nook sitting in a chair in the dark. "Alice. What's wrong?" Alice turned to Carl. She had tears in her eyes. Carl moved closer to her and knelt down. "What is it? What are you crying about, babe?" he asked.

"Carl," Alice started. She wasn't sure if she could tell him. "I ... I ..." She stopped.

Carl took hold of her hand. "What?"

Alice took a deep breath. "I was inoculated."

Carl stared at her. He let go of her hand. "What? What do you mean you?"

Alice was annoyed he let go of her hand and stood, looking down on her. "I was inoculated, Carl. With Cerebtol."

"Cerebtol? Alice." Carl was confused. He felt like his world was spinning out of control. She must be wrong. "You, what? Are you sure?"

"Yes! Yes! I know what I did, Carl!" Alice turned away and started sobbing.

"Oh, Lord, no." Carl knelt back down and placed his large hands on her shoulders. "Alice." He felt like he was going to vomit. His stomach was churning. "When?"

Alice sniffed and wiped her eyes. "March 16th."

"This year?"

Alice became irate. "Yes, this year!"

"Are you sure?"

"Carl! I have a photographic memory, dammit. Yes, this year. Two hundred and eighty-four days ago. Do you want the time and minutes?" she asked sarcastically.

"No, no. Alice." Carl rubbed her shoulder. "Baby. I'm sorry. I ... how could you do this? Why did ...?"

"I was scared, Carl. My dad died form Alzheimer's. My mom has a diagnosis." Alice was speaking rapidly and not giving Carl a chance to say a word. She had this stored up for so long, and was letting it all out. "It runs in my family and I'm susceptible. I was forgetting things and I thought … The doctor ran some tests and confirmed it's Early Onset."

"No!"

"I thought I could get the jump and stop it from getting worse. The tests all showed it worked. There was nothing to worry about. My doctor even recommended I have it."

"Why didn't you wait?" Carl couldn't see her reasoning. "Why did you jump into it? You knew it was a new product."

"I know. But I was scared because I was forgetting so much stuff."

Carl became angry and stood again. "That's no reason to take something like that and not even discuss it with me. I helped create it! Don't you think we should have at least talked about it before you went off and …?"

Alice started sobbing. "Carl. I'm scared," she whimpered as she grabbed his wrist. Carl looked at the fear in her eyes. He squatted onto one knee and brushed her hair back, then wiped her tears.

"I'm sorry, Alice. It's no good blaming." He held her hands. "I'm sorry."

Alice sniffed. "I'm the one who should be sorry."

"Look, it was less than a year ago, and we don't really know what it's doing. Imar is here. We have time. We have good minds who will work on this." Carl turned her head so she could see his eyes. "We're scientists. Let's see what's going on, and beat it."

Alice believed him. She knew he was brilliant. She believed he and Imar could do anything. They beat 19Q. They can beat this virus. They were her heroes.

She hoped.

December 27

THE BOARDROOM AT Johns Hopkins was quiet. Margaret Giddings, the board chairman, sat stoically at the head of the table. A round speaker phone sat in the center of the long, wooden table surrounded by some paper cups, a partially empty box of cookies, and a small stack of notepads. Members of the board of Johns Hopkins University sat around the table in the leather chairs. Only three chairs were empty. Carl Kruger sat at the head of the table opposite the chairman. Imar Spaan sat to Carl's right. His hands were folded on the table.

A projection screen behind Carl displayed the ominous statistics in a columnar format; ninety percent of the Phase One test group were experiencing some sort of memory loss or confusion.

No one said a word. It was eerily quiet.

The speaker phone crackled. "Ninety percent, Doctor?"

Carl cleared his throat. "Yes, ninety. Nine zero."

"And how long since their inoculation?" Margaret asked.

"Just over three years," Carl answered. "We received FDA approval to proceed with Phase One September, 2015. The participants were inoculated October sixth thru October twenty-eighth."

"I don't think we have a choice here, Margaret," one of the board members said. "We should contact the FDA immediately and activate the research team."

"I agree," another said. "The inoculations should stop right away before we cause more damage."

"I don't, gentlemen," Margaret replied. "We do not know if the virus is the cause or not. It could be something completely foreign; vascular dementia brought on by a different catalyst. We don't know, and to suppose our cure is the cause is premature."

"But to have all of them within a short time-frame is too much of a coincidence," Carl said.

"But a coincidence none the less until proven otherwise," the chairman retorted.

"But this could potentially affect millions of people," another member said.

"What is done is done," Margaret replied. "We do not know for certain this is the cause. What we do know is that this is the cure. This drug has improved the lives of many people, even if just on a temporary basis of several years. We have waivers in place. It is still a good cure and the single most profitable development for the University. Until we have evidence to suggest otherwise, I believe we should continue down the path we are on."

Imar leaned forward. "Then, Madam Chairman, this is about money?" he asked.

The woman cleared her throat and also leaned forward. "Mr. Spaan. It is *always* about money."

"Even if it is a looming catastrophe?" Carl asked.

"Again, Doctor Kruger. We do not know that. You need to recall your Nobel Prize team and find out what is going on with our Phase One recipients before we can make an educated determination."

"And the FDA?" came the voice over the phone. "Should we at least alert them to the possibility?"

"No. I don't believe so. Not yet," she replied.

Frustrated, Carl folded his hands and looked at the lady. "I can assemble the team within a few days. I'd like to have additional scientists join the team, particularly a Doctor Norman Raynould." Imar smiled as Carl continued. "He worked on the 19Q project with me. He is a brilliant microbiologist. He won the Nobel in 2003."

"You may have anyone you like on your team, Doctor," Margaret said. "Just find out what is happening and, when you do," the elder woman leaned forward and looked at Carl intently, and continued, "... tell this board first before you tell anyone else. Is that clear, Doctor?"

"It is."

"Good."

Carl cleared his throat. "I understand Doctor Pitke is on vacation."

"He is?" replied Margaret.

"Do we know where he is?" Imar asked.

"No. He left no word with me," Margaret said, showing her disappointment that she would be expected to know his vacation plans. "I'm not his secretary. I suggest you contact her about his schedule," she said sarcastically.

"Yes, Madam Chairman," Carl replied.

Margaret smiled. "By the way, doctors. Congratulations on your prestigious award."

Carl looked at Imar and back to the woman. She had an almost sinister look on her wrinkled face. "Thank you," he said quietly as he gathered his papers to leave.

The Memory Project was reactivated.

—*—

The drive home from the university was quiet. Imar said hardly a word as Carl drove through the sleet and rain. It was cold, icy and dangerous outside, so he had to keep his speed down even though the roads were sanded. Imar watched as Carl drove down the almost empty road toward the house.

"I'm going to call Norman as soon as we get back to the house," Carl said.

"Good idea," Imar replied. "Tell him I said 'Hi'."

"I don't understand how they can negate the seriousness of this, Imar."

"Nor do I, Carl. Nor do I."

"The patients who took this will be in a much worse state than ever. Doesn't she get that?"

"I suppose not."

They pulled up to the house. Imar said his pleasantries as he went upstairs to bed. Carl went into the study and called Norman Raynould. "Hello?" came the crackly voice over the phone.

"Norm. This is Carl Kruger."

"Carl? Why, I would've never ... Carl?"

"Yes, Carl Kruger."

"The Nobel Prize recipient in Medicine Carl Kruger?"

"Yes, Norm."

"It is good to hear your voice, my friend."

Carl paused. "As is yours, Norm. But I have a problem. A big problem."

"Problem?"

"Yes. I need your help."

"It's that chemical element we found in the virus, isn't it?"

Carl was taken aback for a second. "I don't know, but there is something wrong, Norm. The Phase One recipients are relapsing."

"Oh, no."

"Yes. Ninety percent are experiencing some form of memory deficiency."

"Ninety?"

"Yes. Norm. I need your help. I need to find out why, right away. Can you come out?"

"Sure. I can be there by ... hold on a second. Okay. Yeah. I can be there by Monday."

"That's New Year's eve."

"Yes, it is. I look forward to celebrating with my old friend."

"Friends," Carl corrected him. "Imar is here."

Norman laughed. "Well, then. I look forward to seeing that Einstein impersonator."

"I can put you up in the house if you don't mind sharing a room with Einstein. It has twin beds, a TV ..."

"No. No thank you. He's brilliant, but he snores like a water buffalo. I'll stay at a hotel."

Carl laughed. "Thank you, Norm."

"Carl. No thanks needed. You will get my bill later."

Carl laughed. "Okay, then. Until Monday. Travel safe."

"I will."

Carl felt better as he hung up the phone. Then, he realized he had to call all of the key scientists and doctors back to the university to find out why their miracle drug was failing. It was late; he was tired. He may as well get some sleep and start the calls tomorrow.

December 31

IT WAS NEW Year's Eve. Carl and Imar had just picked up Norman Raynould at the airport. The three friends acted like college buddies at a reunion when they saw each other at the terminal. Norman waved when he saw his friends. Imar threw his hands up and squealed like a little girl. Carl laughed at the two of them; brilliant scientists making a mockery of their meeting for Carl's entertainment. It was good to be with them. It was a shame that it had to be for all the wrong reasons.

Carl spent the past three days going over more emails and phone messages. Many were from some of the same people who left messages before. There seemed to be a little more panic in each voicemail they left. Others seemed to be more apathetic. All were worried. Carl set the systems up for standard responses that said the department was closed and their inquiries would be addressed on January 3rd, right after the holidays. That would give him and his team a little time to develop some talking points for the numerous calls that he expected after their return.

The majority of his team planned to be onsite on January 2nd. Carl planned to brief them on the current situation and stress the importance of confidentiality until they could determine what the problem was. He knew his team would adhere to that. They were experts, professionals and heavily vested in the participation and success of the product. Even though Carl and Imar took the honors at the Nobel, they were all acknowledged for their contribution and success. It was a major shot in the arm for many of their careers.

Most of them planned to show up early, about six AM, and get their stations organized. Carl planned a staff meeting at eight. He felt prepared and somewhat in control of a volatile situation. It would take some doing getting the secure area and labs back. When the Memory Project was scaled back, JHU turned most of the labs and facilities over to the team studying the West Nile Virus. Now they would have to relocate to another lab on campus and allow the Memory Project to return.

What puzzled Carl the most was Owen. He was nowhere to be found. Carl contacted Owen's secretary who said she thought he was going to be back before New Year's. Carl asked her if he had travel plans and he didn't. She said he was going to spend a quiet holiday at home, read and maybe go to Atlantic City for a few days and hit the tables. Otherwise, no travel plans that she was aware of.

Carl called Owen's cell phone and continued to get a full voice mailbox. He searched his desk and found notes, plans and calendar appointments indicating he planned to do some work on old projects, purge some files and get ready for another year, but nothing about travel plans. Everything looked normal. Yet, it looked all wrong.

Carl even took the time one afternoon to drive by Owen's house in Thurmont near Cunningham Falls State Park. He could see why Owen lived so far out of town. The drive was beautiful even during the winter. Owen said he had to rise early to beat the traffic, but it was worth it. Now Carl knew why.

The driveway to Owen's house was lined with barren birch trees. The house sat a couple of hundred feet back off the main road. Carl pulled into the driveway and stopped. The two feet of snow that fell three days ago was still packed on the driveway. It made the approach in Carl's car impossible. He knew no one had been in or out of the house since then. No reason to go to the house. If Owen was there, he was holed up for a reason, or dead. Carl knew he was gone. The numerous emails and voice mail messages, the look Owen gave him at the airport, and the small salute which haunted him, was all Carl needed to convince himself Owen knew what was coming down and left.

—*—

It was almost midnight. Owen sat in his overstuffed chair wearing a heavy robe and warm slippers. It was cold outside, but he was warm in his luxurious room. The television was playing clips from various cities around the world celebrating the coming year. The fireworks over the Sydney Opera House, Tokyo and Beijing were spectacular. Owen watched the time countdown on the clock in Delhi, India on TV. When it struck midnight, he lifted his glass of Scotch. He sloshed some that spilled on his robe. He was already drunk and didn't care.

"Happy New Year, you fools," he said as he drained the last of the alcohol. He rested his arm over the edge of the chair and, as he dozed off to sleep, he slowly let go of the glass which fell to the floor and broke. He didn't wake.

—*—

The Kruger house was buzzing just a few minutes before midnight. Alice and Carl had planned long ago to have several guests over to celebrate the New Year and Carl's Nobel Prize. Alice even promised Joey he could have some friends for a sleepover and they could stay up until midnight. She said they could bang pans outside when the clock struck twelve, which was why Joey was franticly looking for a big pan and spoon. It was just a few minutes before midnight, so he had to hurry.

Imar and Norman were drinking some champagne and having a rather heated discussion about using palladium as the source for restructuring the hydrogen and oxygen atoms from water to create a new unlimited energy source. Guests observed their brilliance, though both were slightly inebriated. Their passion for knowledge never waned.

Alice and Carl were standing near the fireplace observing their guests. Carl had a small glass of wine. Alice was drinking sparkling cider. Joey and three boys ran by them with pans and metal spoons. "Wait!" Alice yelled. "Use the wooden spoons."

The boys did an about face and headed back into the kitchen to get wooden spoons. Carl and Alice laughed as the foursome ran past them again, heading through the crowd to the front door. "Wait until you hear us cheer," Carl yelled as they flew out the door into the cold, night air. "Ahhh. To be young, again," he said.

"It's almost midnight," Alice said. "Carl. This has been a great evening. Thank you for helping me take my mind off … off …"

Carl leaned forward and touched noses with her. "You're welcome, baby," he said. His deep voice was soothing, reassuring to Alice.

"One minute to midnight," someone said across the room. "Fill your glasses," someone else said. "Find your partner," another person said.

Imar and Norman stopped their conversation and looked at each other, then turned and walked in opposite directions. "Ten, nine, eight …" The crowd started the countdown. "… three, two, one. Happy New Year," the people said in unison. Most people raised their glasses, others kissed their partner. Carl and Alice kissed, toasted, and sipped their drinks.

Imar walked over to celebrate. "Happy New Year," he said. Alice kissed him on the cheek.

Norman walked up and saw the kiss. "Hey. I could use one, too," he said.

Imar leaned over and kissed Norman on the cheek. "There!"

Norman quickly wiped the kiss off. "Not from you, you crazy old scientist." Alice leaned forward and kissed Norman on the cheek. "That's better," he said.

The four of them laughed heartily. They lifted their glasses together and celebrated the New Year.

They had no idea what lay ahead.

Year Six—2019

Square One

January 2

IT WAS EIGHT o'clock in the morning at the Alzheimer's Disease Research Center at Johns Hopkins University. Carl believed the cafeteria was the best area to have the team meeting. People were sitting at tables with coffee, rolls and fruit talking about the rumors they heard about the virus. Carl, Imar and Norman were standing at the back of the room with a large whiteboard which had series of columns and numbers scribbled on it.

"Folks, if I can have your attention, please," Carl started. The room quickly quieted down. "Thank you for coming in this morning. The holidays are over, and we have some troubling news. But first, I need to make very clear that anything shared with you from this point on, absolutely cannot be shared with anyone outside of this team; not even your spouses. No one." The people looked around and a couple whispered. "We have unverified information that some participants in the Phase One group of Cerebtol are having a relapse." A couple of people gasped. "I know. This is upsetting. We don't know what is occurring, but we will find out. I returned from my vacation to find emails and voicemails from participants and their family members that people were experiencing memory lapses. It appears from the information so far, there is no distinction from age or sex for the lapse. Nearly all of the participants are experiencing some type …"

"All?" came a question from the audience.

"Yes. All. Something is happening and we have to find out why. It could be a vascular event brought on by a foreign catalyst interacting with the virus, or a chemical. We don't know. All we know so far is there were multiple emails and voice mails left here over the last six weeks from people, some panicked, about experiencing a relapse of memory deficits. We also know the frequency of the calls and contacts are increasing." Carl observed the audience. They were stunned. No one said a word. All eyes were on him. "We need to find out why and stop it."

"When did it start?" came another question.

"Around November, from what I can decipher. Maybe longer. I was in Sweden …" Carl paused for a second, remembering his moment of glory that now seemed so fleeting, "… when some of the calls came into the center. Subsequent calls and contacts were forwarded to me by Doctor Pitke."

"Where is he?" someone asked.

Carl cleared his throat. "Uh, I don't know." He could see the puzzled look on everyone's face, and continued. "We are trying to locate him. He might be on an extended vacation, but no contact information was left with his secretary or staff when he left."

"Gone?"

"Yes. We hope that he is alright and just stuck somewhere, maybe in a foreign country on vacation. We will let you know when we find out. For now, I will be in charge of this research project." Carl turned to his right. "This is Doctor Norman Raynould. He is a world renown molecular physicist and Nobel recipient who has worked with me before." Norman bowed slightly. "And, of course, many of you remember Doctor Spaan." Imar waved and smiled.

"Where do we start, Doctor?" came another question.

"We need data," Carl said. "Samples. Tests. Information. Don, I need you to create a team to test some of the patients."

A middle-aged man slightly balding replied. "We can fly them here and conduct the tests."

"Or you can go there," Carl said. "We need information now. Whatever it takes." He looked to a young lady in the front row. "Mary Beth, I need you to exhume the bodies of any patients who have expired." The lady nodded. "I believe we only have three, but check the current status." Carl pointed to Imar. "Imar will oversee any brain biopsies that we can perform." Imar nodded. "Norman will lead the sample analysis." Norman nodded.

"Folks, as of this moment, the Memory Project is reactivated. We have to do this as fast as we can. There is a reason this is happening, and we are going to find out what it is, and stop it."

"If we can," came a response from the audience.

Carl confirmed. "Yes, if we can. And with God's help, I believe we will."

January 11

THE SUN WAS shining brightly and reflecting off the sunglasses of the young woman standing near the portable, blue awning. Mary Beth was dressed in shorts, sandals, a short-sleeved top and was wearing a Sun Devils baseball cap. It was a great change for her from cold, Baltimore, Maryland to the land of sunshine; Palm Springs, California. She was standing next to a man wearing a sports shirt and a baseball cap with the words 'Medical Examiner, Palm Springs' written across the front.

Black smoke billowed from the exhaust of the backhoe as it revved its engine to dig into the hard, sun-baked ground. "Hard as rock, lady. Hard as rock," the chubby, dirty driver yelled from the cab of the backhoe. His beard looked matted and greasy.

Mary Beth watched expressionless. "Just keep digging," she yelled back. "I need that body right away."

Two other men dressed in coveralls watched as the backhoe moved shovel after shovel of dirt from the area and placed it in a large pile. One of the men pulled out a tape measure and checked the depth of the hole. "That's it," he yelled, and the backhoe operator turned the motor off and climbed down from the cab.

"Now it's by hand," he said to the woman, who could care less about the unfolding events. She was typing furiously on her PED.

The three men used shovels to remove layer after layer of dirt until they came to a hard object. They slowly uncovered the object to reveal a brown, metal casket in the dirt. "O.K. Let's hook her up," one guy yelled to the others.

The backhoe driver climbed back into the cab with great effort and lowered the bucket. Two men hooked chains around the bucket and connected it to the appendages on the sides of the casket. "Okay!" one guy said.

The backhoe slowly lifted the casket out of the ground. Dirt fell from the top and sides as the operator swung the casket around and onto a large cart attached to a golf cart. The men unhooked the chains and threw them onto the cart next to the casket.

The backhoe driver climbed down from the cab and walked over to the woman and man who were watching the men dig up the casket. "Say hello to Ms. Emily Brownlee," the backhoe operator said as he motioned to the casket.

"Cute," Mary Beth said, unamused.

The backhoe operator signed a document and handed it to the man, who signed the document and gave a copy to the backhoe operator. "Thanks. If you could have her back here by curfew, I'd appreciate it," the backhoe operator said with a laugh. He motioned to the two other men to take the casket to the waiting medical examiner's van parked at the curb. The men loaded the casket into the van as the woman and man climbed in. After they shut the door, they drove off through the cemetery with Mary Beth in the back of the van.

—*—

A wealthy man in a poor country doesn't have to wait for anything. So Owen discovered.

In less than a month, the city of Thimphu was able to work with the district in Bhutan to approve the construction of Owen's Shangri-la; a veritable fortress nestled in the hills above the Thimphu Valley in Bhutan. The architect and engineer that Owen hired drafted plans to develop the grounds over three phases. The first phase would include the utilities, first section of the house and security. The second phase would address the second section of the house, the work areas and shops. The third phase would complete the house and guest quarters and agriculture needs.

The black Mercedes pulled up to the parking area where several men in wool coats stood looking over some blueprints spread out on the tailgate of a truck. A bulldozer was busy in the background moving mounds of dirt to level the area where the house was going to be situated. Black smoke billowed as the dozer struggled to break through the rock and hard soil. The driver of the car stepped out and opened the door. Owen climbed out and casually surveyed the scene. He had some documents under his arm as he walked toward the men by the truck.

"Good morning, Meester Wallace," one of the men greeted. "As you can see, sir, we have begun to begin the work."

Idiot, Owen thought. He looked at the plans on the tailgate. "How long until I can move into the first section of the house?" The men looked at each other. Owen continued. "Look, I don't want to live in that hotel forever. How long?"

"About three months, Meester Wallace," one of the men said.

Owen smiled. "Make it one month, and I'll double the first installment."

The men looked at each other and spoke in their native dialects. The engineer spoke for the group. "Very well, Meester Wallace. One month; double the money." Owen reached out and shook his hand. "Good. The sooner the better," he said. "I'm ready to get out of that hotel and stretch my legs."

The men looked at him curiously, then at his legs, then back to him.

Idiots, Owen thought as he turned and got back into his Mercedes.

—*—

Alice dropped Joey off at school and was headed to the natural vitamins store to see what she could get to help prevent her impending memory loss. Each time something came up where she second guessed it, she thought it was her memory. Her anxiety was escalating. She needed something to use to give her hope right now. Maybe if she started taking some natural supplements, it could delay the process and give Carl and his team time to find a cure.

The store was large and well stocked of almost every natural herb and supplement known. Alice wandered through the aisles looking at the categories; muscle building, joints, digestion, memory and brain. She was looking at the items on the long shelf when an employee walked up. "May I help you find something?" The employee was a young man, maybe twenty, well groomed and fit. He had an infectious smile, slightly spiked hair, and a nice tan.

"Oh. I don't know. I was looking for natural memory improvement products and was just ..." Alice started.

"Oh. Well, here. We have a huge assortment of natural supplements to aid with memory functions, cognition, neurogenesis ..."

"Neurogenesis? Are you kidding?"

"No. Not at all." The young man reached over and grabbed a bottle off the shelf. "This is PS9, a plant derivative that was recently discovered in the Amazon jungles. It contains trace elements of pregnenolone sulfate; a steroid that is associated with neurogenesis of the hippocampus."

Alice took the bottle and read the label. "This is amazing."

"Yes," he replied. "We have several other aids to memory; Lemon balm with GABA transaminase inhibitor properties, Theanine green tea which increases brain serotonin and dopamine levels, grape seed extract, ginsing, kava ..." He pointed to each item as he named them with their associated properties.

Alice was stunned. "I had no idea."

The young man smiled. "Many people don't."

"There are so many. I don't know where to start."

"I suggest you get the PS9, lemon balm, and Theanine green tea to start."

"Okay. I think I will. But I'm still going to do some research on this before I take them." Alice took a bottle of each from the young man. "Thank you," she said as she headed to the counter to check out.

—*—

The elderly man was helped into the chair by his daughter; a middle-aged woman with auburn hair. "Here, Dad." The doctor and a nurse's assistant were standing by the chair preparing some electrode attachments and such. "They're just going to run the same tests on you that they did a couple of years ago," the daughter said reassuringly.

"I know that. I haven't completely lost my mind, you know."

The daughter turned to the doctor. "He's been a little agitated lately," she whispered.

"And I certainly haven't lost my hearing!" the elderly man yelled.

"That's O.K., Mr. Clarkson. We'll get you out of here in a jiffy," the young, male doctor said. "Right after we run these tests."

The doctor hooked the electrodes to various areas on the scalp and started the scan. The assistant turned a few dials and images began appearing on the screen. The brain areas were mapped in blue with touches of yellow and red. The doctor began speaking words to the older man as his assistant captured images through the keyboard. "Ok. Here we go." The doctor waited a few seconds, then said "Hot."

Click.

"Happy."

Click.

"Lost."

Click.

"Cold."

Click.

This continued for several minutes as the doctors mapped the man's brain activity for the university. "There," the doctor said. "That didn't take long."

"For you," the old man retorted.

"Now, Mr. Clarkson, I'll need to have you go to 1C for your MRI, then that should do it for the day," the assistant said.

"Great. Another test," the old man replied. "Makes me feel like a rat."

"It's temporary, Dad. They just need to see what's going on up there," his daughter said as she pointed to his head.

The old man stopped and glared at her. "I know what's going on, Lucy. I'm losing my marbles again," the old man said as he turned away and headed out the door to 1C; wherever that was.

—*—

The elderly woman sat in the chair trying to see the halo around her head with the wires hanging off. Her husband sat next to her holding her hand. "It will be alright, honey," he said reassuringly. The woman never said a word, but continued to visually trace the wire leads to the control panel.

The headring for the Stereotactic Brain Biopsy was mounted to her skull with pins. It didn't hurt because the doctor numbed the area before placing the ring on her head. It was critical that she stay still, so the doctor asked her husband to come in, hold her hand, and try to keep her calm.

The doctor made a small incision in the numbed area of the scalp. He started the drill and prepared to enter the skull. The woman squeezed her husband's hand. "It's okay, honey. It will only take a few minutes. You won't feel a thing. I won't let them hurt you," he said as tears formed in his eyes.

The woman could feel the pressure as the drill bit penetrated the hard bone with a high pitch whine. It lasted for a few seconds and stopped. The woman relaxed her grip. "See? I told you it wouldn't hurt," her husband said.

"You'll feel a little pressure now," the doctor said as he slowly inserted the needle into the brain. The woman jerked slightly and the doctor stopped. "Please try and be still, Lola," he said. He continued to push the needle into the brain, and extract the sample. "There. Finished."

The elderly man kissed his wife on the forehead.

—*—

"Everything?" the young man asked as he adjusted his glasses.

Carl was direct with his response. "Yes, everything. I want every piece of data searched on this station and in his directories."

"I don't want to anger you, but I don't think you can request a data search of your boss's computer files and activities. I think I need to get approval from a higher authority, Doctor."

"Like who?" Carl asked.

"Like … his boss? I think that would be the Dean of Research and Engineering."

"I know who it is. How about the Universty Board Chairman? Is that high enough?" Carl asked cynically.

The young man was embarrassed. "Yes, doctor. It would be."

"Then you'll have it by tomorrow. I want you to rip this apart and tell me what has been deleted in the past six months."

"That's going to take some time," the young man answered.

"We don't have very much of that, I'm afraid," Carl replied.

January 13

"HERE THEY COME now." Imar was almost giddy watching the decontamination chamber blow debris off the analyst standing in the center. She was holding a cooler in her hand as the air blew her hair and clothes, almost blowing her glasses off her nose.

"The samples?" Carl asked.

"Yes. The brain tissue samples from the biopsies that were performed yesterday." Imar walked over to the door as it opened. "Thank you, Janice. I'll take those from here." The young lady handed Imar the cooler and walked across the room to her station. Imar took the cooler and placed it inside the glass isolation chamber sitting on a stand in the center of the room.

"I'll stop back later, Imar. I can see you have your work cut out for you," Carl said as he started toward the exit door to the decontamination chamber.

"O.K.," Imar said as he closed the door to the glass box. He didn't take his eyes off the cooler.

Carl went through the decontamination blower and out the two doors to the center entrance. The computer technician was standing at the entrance talking with the security guard when Carl appeared. "Here he is," the guard said.

"Doctor Kruger. I have some of the information, sir," the technician said as Carl approached him. He adjusted his glasses and handed Carl a couple of pieces of paper.

"So, you must have received the approval from Ms. Giddings," Carl said.

"Who?"

"The chairman of the board."

"No, sir. I received a call from the VP of IT. He told me to get on this project," the young man said adjusting his glasses again.

Carl smiled. "Of course. What do you have?"

Carl glanced at the documents as the young man began to speak. "Mr. Pitke …"

"Hold it!" Carl said to stop the boy. "Let's move this to the hallway." He motioned for the two of them to go into the hallway out of earshot from the security guard. "Now, you were saying?"

"Mr. Pitke deleted some emails, mostly those received after the first of November."

"Were there some before that time?" Carl asked.

"Yes, but we haven't reconstructed all of them yet."

"And the nature of the emails?"

"Right here, sir," the boy said as he pointed to the paper in Carl's hand. Carl scanned it and then looked at the boy. The boy cleared his throat. "We have more work to do, Doctor Kruger," the boy said as he adjusted his glasses.

"Yes, you do. And no one is to know about this, do you understand?" Carl said.

"I do, sir."

Carl leaned closer to the boy. His deep voice and size was quite intimidating to the young man. "Do you? Because this is a serious, serious matter that cannot leak out to anyone."

The young man started to shake slightly. "Yes, sir. Understood. You have my word."

"And I don't want anyone else working on this except you. If you need help, you let me know; not your VP. Is that clear?"

"Yes, sir. Very clear."

"Good." Carl folded the papers and placed them inside his jacket and started to walk off. He stopped and went back to the young man who was still frozen in place. "Thank you … uh"

"Kelly, sir. Kelly Winters."

"Kelly. Thank you," Carl said sincerely.

The young man smiled as Carl turned and walked down the hallway and out of sight.

—*—

Jamille Larson was sitting down to dinner at the Delta Queen with a couple of friends; DeShawn, his basketball buddy, his wife Carole, and DeShawn's sister Marie. Carole thought Marie would be a perfect match for Jamille, so she talked DeShawn into arranging the double date in the hopes of a successful matchmaking event. Jamille knew what Carole had planned. It wasn't the first time she tried to match him with a friend. But one of his friend's sisters? That seemed to be over the top for Jamille. He wasn't very interested, but was always polite. He saw tonight as just another night for a good meal, light banter and a parting of ways.

The group was enjoying a few glasses of wine and appetizers. They were laughing and telling short stories and jokes. Everyone was in a good mood. The four of them were enjoying themselves and the company, especially Jamille and Marie.

The main course had just been served when Jamille's phone rang. He looked at his phone and saw that the call was from Carl. "Uh, I'm sorry. Will you please excuse me?"

"No," Carole said.

"Pardon?" Jamille asked with a surprised look. The reply caught him off guard.

"No! You always get called away when we're just starting dinner and getting to know each other," Carole said.

"Oh, so this happens a lot?" Marie asked. The atmosphere quickly went sour. "You mean there are other dates and I'm just another fill-in? A last resort?"

"No, no, sis." DeShawn could see his sister was instantly offended at being a fill-in double date and he tried in vain to smooth things over. "You're not a last resort. There's other girls we could of called." He realized what he said as the words left his mouth. "I mean …"

"What?" Marie yelped. "You what?"

Jamille lifted the phone to his ear. "Hello?" he said as he stood and turned away from the table trying to listen to the voice over the phone.

"How can you say that, DeShawn?" Carole said.

Jamille put his finger in the other ear to block out the escalating argument going on at his table as he moved farther and farther away. "Yeah. Carl? Yeah. It's Jamille. What? Problem?" Jamille continued to scoot along the wall until he came to an exit to the deck of the riverboat. He stepped outside into the quiet, cool air and listened to Carl intently. "Yeah, I guess I can. I can request a temporary assignment to Baltimore. We do it all the time. Sure." Jamille looked through the window into the restaurant as Marie grabbed her purse and stormed out of the restaurant, knocking over a glass of wine on the table. Carole shook her finger at DeShawn and threw her napkin in his face. She grabbed her sweater and followed her friend out the door. DeShawn shook his head and watched the ladies leave in anger. He glanced over to Jamille who was still outside the restaurant talking on his cell phone. DeShawn filled his glass with wine, raised it to Jamille and drained it in one huge gulp. "Yeah," Jamille continued on the phone watching DeShawn inhale the wine. "I can leave sometime tomorrow. There's nothing holding me here. That's for sure."

January 22

THE RESEARCH CENTER had been busy like never before. Many people were working sixteen to eighteen hours a day trying to find the reason for the memory loss of the Phase One patients. Norman Raynould was receiving samples from all over the country. He assured the samples would be safe in the lab which was restored to a level four biohazard facility with maximum security protocols. The retinal, palm and voice recognition systems were reactivated. Guards were posted everywhere. Norman divided the samples into "active" for live patients, and "inactive" for deceased. Staff was already analyzing the "inactive" samples and making comparisons as they came in. Only two people in the test group had died from natural causes and their bodies exhumed. Samples of their brains, tissues, and spinal fluid were taken from their bodies and shipped to the lab in a couple of days. Mary Beth did a great job of executing the court order for the exhumation. She got a nice tan, too.

Thirty-two people of the Phase One test group cooperated with additional testing for cognition and memory. They were the same tests they had done before; CRI, SPECT, MRI and PET scans were performed on as many patients as they could in the short time. Don Willows, the resident Doctor of Physiology at the university, personally accompanied as many tests as he could to observe performance and behavior. The scans were sent to the university for follow-up and baseline comparisons. The scores from the cognition and memory tests were categorized into percentile rankings for the group as a whole. It was clear the entire test group was dropping in memory. The group as a whole dropped seven percent in both. Some individuals were experiencing a ten percent reduction in memory. Patients previously diagnosed with Early Onset Alzheimer's dropped fifteen percent in memory retention.

The six samples from the brain biopsies were more than anyone imagined receiving. Imar Spaan and his team worked steadily through the past week analyzing the tissue samples and comparing them to prior biopsies of other patients. All six samples were in a secure storage container behind a steel door with a window in it. Imar was walking by the unit when he stopped and looked in to see a container labeled 'March 5, 2015—MacKenna, S.' Imar smiled and whispered, "Thank you, Mac," as he continued down the hall toward the exit carrying a pile of documents.

Carl and Norman were in the main conference room waiting for Imar. Norman had his notes in a stack on the table in front of him. He handed Carl a summary of the results of the testing that he managed. He had a dual overhead display set up to show test comparisons of the Phase One patients to baseline and the population group, and a second screen for other comparisons and notes. The left screen displayed images of PET scans of a brain before the inoculation, at the end of Phase One testing, and last week. The most recent scan showed obvious deterioration of brain activity from the end of Phase One, but not as bad as when first inoculated. The second screen was blank.

Imar walked in and set his documents on the table. He looked at the screen and saw the images of brain activity. He pointed to one of the images. "Current?" he asked.

"Yes," Norman replied.

"I see," Imar said as he motioned for the keyboard. "May I?"

"Sure," Norman said as he scooted the keyboard toward Imar. Imar typed in a few keystrokes and an image popped up on the blank screen. Carl and Norman watched as the image displayed little circles with tentacles connecting to each other, creating a mass of red.

"Gentlemen, this is a simulation of the JC2 virus that we have used in Cerebtol," Imar started.

"Yes, I recognize the image," Carl said.

The image zoomed out to show the virus cluster inside of a brain cell, connected to another cell. As the image continued to zoom out, soon a cluster of cells came into focus in the form of tangles. Then impulses simulated as little flashes of electrical current jumped from cell to cell, then area to area.

"The virus clustered inside damaged brain cells and revitalized them as a surrogate cell, repairing the damaged cells and neurofibrillary tangles of the microtubule structures, thus allowing neurotransmission to occur."

"Yes," Norman replied.

"This is what's happening now." Imar tapped the keyboard. The electrical impulses slowed and soon stopped. The tangles withered and the microtubule structures began collapsing. The affected area was depicted in a shade of gray that spread slowly across the brain surface. As the image zoomed in, it displayed a dead brain cell, then withered viruses inside the cell. "The virus is killing the cell."

"So, it *is* the virus after all," Carl said dejectedly.

"I'm afraid it is," Imar replied.

"Do we know why?" Norman asked.

"No. Not yet. It appears there may be a deterioration of the protein shell, or maybe the enzyme. We don't know yet," Imar answered.

The display continued to show the creation and expansion of the virus on one side, and the deterioration and death of the brain on the other.

Over and over again.

The three men stared at the images.

Carl spoke first. "You've double checked your findings?"

"Triple checked. And we are still checking, Carl. But, the reality is, this is what is now happening and will happen to the millions of people who took Cerebtol."

"Everyone?" came a question from across the room. Alice was standing by the doorway looking at the men. They didn't hear her approach even though she was using her crutches. She had a look of terror on her face.

"Alice!" Carl said as he stood to his feet.

"I asked, everyone?"

Imar cleared his throat. "Uh, yes, Alice. I expect everyone."

Alice turned and hobbled away with Carl walking quickly to catch up to her.

—*—

It was late. Carl was tired. It was undoubtedly one of the worst days he's ever had. Alice overheard him, Imar and Norman discussing the reversion of the JC2 virus. She wanted to be left alone, but Carl couldn't leave her in that condition. She was distraught thinking she would eventually become a vegetable, a shell of a human being like many advanced Alzheimer's patients do. Waves of fear rolled over her every few minutes. She was a brilliant scientist. She wanted to do something to try and stop this thing, but she didn't know how or where to even start. She felt helpless and alone, even with Carl near her. No one knew what she was feeling, and she couldn't express it. She didn't want to lose her memories. She had so many good memories. Each was a treasure.

Carl assured her there was time. It was almost three years from the time of inoculation to the time of initial decompensation. That would mean she would have another two years before she would start to experience any symptoms. That would give them plenty of time to find a way to stop the deterioration. Carl was confident of that. Alice wasn't. He was working with some of the greatest minds in the world. Besides, it would only kill the brain cells that were already damaged, so she would not be in a worse state; just where she was when she started. He tried to convince Alice there was hope. If only he could convince himself.

"Doctor Kruger," Margaret Giddings said as she approached the table.

Carl was sitting alone at the restaurant waiting for his cup of coffee. "Madam, Chair," he said as he stood while she seated herself. "Thank you for meeting me on such short notice."

"I am not sure I had any choice, Doctor," she said with a wry smile. "You said it was urgent and that we needed to meet tonight, without the board."

"Yes, ma'am," he replied.

"We may very well be acting outside of proper protocol if this deals with university issues, Doctor." Margaret turned to the waiter who walked up to her side with a menu. She waved off the menu and said, "I'd like some hot, raspberry tea with lemon, please. That's all." She turned back to Carl not giving the waiter an opportunity to reply nor take an order from Carl. "I'm not sure I will be of any help to you, Doctor Kruger."

"Madam Chairman, we have confirmed our assumptions." Carl paused, waiting for a reply from the woman who seemed oblivious. "The virus is breaking down in Phase One participants."

"I see."

The two sat quietly at the table. Carl offered nothing further. Margaret required nothing more. The elderly woman broke the silence. "And what are you going to do?"

"To be honest, I don't know."

"Tea, ma'am," the waiter said as he placed the teapot, cup and saucer on the table. "Here's sweetener, sugar, and some lemon. Can I get …?"

"No, thank you," she said tersely as she waved the waiter off and turned back to Carl. "And what do you want from me or the board now, Doctor?"

Carl cleared his throat. "First, I would like to thank you for authorizing the search of Owen Pitke's files."

"Is he still missing?" she asked with a chuckle.

"Yes, he is. We found more information about this … this … 'event' in those files."

"You think he was covering up something?"

"Yes, ma'am. I do," Carl replied.

"Huh. Do say."

"I believe we should contact the FDA and put them on notice about Cerebtol."

"Why so soon, Doctor? Shouldn't we give some time to see if a cure can be developed before we raise the alarm?"

"No," Carl said emphatically. It startled Margaret. "Too many people are already at risk. We can't allow the inoculations to continue."

Margaret thought for a moment and sipped her tea. "I agree. However, I believe there is a way to inform the FDA without causing widespread panic. We should stop future inoculations, but not reveal the current issue."

"We can't do that! People need to know …"

"Doctor Kruger," she whispered. "May I remind you we are in a public place and you should control your tone and volume, sir."

Carl took a deep breath. "We need to let the people know."

"It would serve no purpose, Doctor."

"What do you mean? How can you say that?" Carl asked.

Margaret smiled and sipped her tea. "If I understand the function of this virus, if you create a way to kill it, then the host cells will die along with it. Either way, the cells are going to die. They were damaged to begin with, so they would return to their dysfunctional state. I believe it is better to allow the process to proceed naturally than to expedite their demise." She took another sip of tea. "At least then you are enabling people to live a better life for a period of time rather than immediately returning them to their hell." She smiled. "It is still a good product, even though it is temporary. Besides, that should give you some time to, hopefully, find an alternative."

Carl took a sip of water and thought. She was right. There was still some good in allowing the virus to progress naturally, even though temporary, than to stop it completely. "Then, I will contact the FDA tomorrow and have them issue a notice that we are voluntarily …"

"No, no, dear Doctor," Margaret said condescendingly. "I will contact Timore Pharmaceuticals and inform them they must cease the distribution of the product."

"Distribution? What about production?"

"They can produce all they like, Doctor. I expect you will find a solution for this dilemma and we will be back in the game, so to speak. If not, then we can certainly package it as a temporary benefit that slows the progression of AD and like diseases for several years. That in itself is still a major breakthrough."

"But there's also Antrole Pharmaceuticals producing the virus and inoculation. What about them?" Carl asked.

"I will give them a call. I'm sure they will also hold off distribution until we know more about the solution."

"That won't fix the product already in the distribution channel, Madam Chairman," Carl said.

"I am certain the pharmaceutical companies will find a way to recall the batches in the channel. It happens all of the time with flu vaccinations, so this won't be any different."

Carl pondered the situation. It seemed like she had all of the answers and they made sense. But deep down, he felt like this was wrong; all wrong. "I don't know. I guess I'm O.K. with that."

Margaret leaned forward and whispered ever so softly. "Doctor, I am not seeking your approval for this," she said as she smiled. She picked up her tea and sipped it leisurely.

January 25

FDA NEWS RELEASE

For Immediate Release: January 25, 2019
Media Inquiries: Jennifer Sands, 301-555-4666,
jennifer.sands@fda.hhs.gov
Consumer Inquiries: 888-INFO-FDA

Timore Pharmaceuticals, Inc. has instituted a voluntary recall of the vaccine Cerebtol, a drug used to prevent dementia in people with either Alzheimer's Disease or Early Onset Alzheimer's Disease

The U.S. Food and Drug Administration today was informed by Timore Pharmaceuticals, Inc., of a possible contamination of the Cerebtol vaccine manufactured after March 25, 2018. The company directs anyone in possession of the vaccine to please contact *Guardian Biopharma Storage Services at 888-555-2233* for return instructions.

Timore Pharmaceuticals, Inc., has guaranteed a full refund to the returning entity.

Cerebtol is marketed in the United States by Timore Pharmaceuticals, Inc., a recipient of and partner with the Howard Foundation of Johns Hopkins University.

On June 20, 2017, the FDA approved Accelerated Development
Review of Cerebtol to reduce and reverse the effects of
Alzheimer's Disease.

—*—

The main building of Shangri-la was taking shape in just two weeks. Owen
was intrigued by a country that employed some of the latest technology
with some of the oldest. Heavy equipment moved mountains of dirt to create a long driveway to the house while local laborers used donkeys and
llamas with baskets to move rocks and debris from the garden areas.

The dozer cleared trees, rocks and dirt from a strip around the hillsides
to provide a clear view of any approaching vehicle or person. The back of
the property abutted a cliff that dropped two hundred feet to the Thimphu
river. The view of the Thimphu valley below down to the city lights was
impressive. Farther upstream, a pipeline was laid to access the pure spring
waters from the river and channel them to the water purification system.
Owen wasn't taking any chances that someone might try to spike his water.

The foreman for the project walked over to greet Owen when he
stepped out of the Mercedes. "It ees looking good, no Meester Wallace?"

Owen looked around. "It is." He pulled his wool coat tight to fend off
the bone-chilling fog.

The foreman got a huge smile. He knew he would likely get a big
bonus from the generous, wealthy man. "Eees dere anyting you would like
to change?" he asked as he laid the plans on the hood of the car.

"Yes. Take your papers off my car, for one," Owen said with a stern
look.

The man quickly removed the papers. "I am sorry, Meester Wallace."

Owen walked over to the edge of the driveway and looked at the progress. The main residence was taking shape. The foundation was laid. The
conduits for the electrical were run to the fence, gate, house and perimeter
lights. The open trenches were ready to be covered by the workers with
carts and shovels.

The foreman walked up and stood next to Owen watching the people work.

"I told you I wanted the main entrance to the house facing south," Owen said, pointing to the foundation. "Isn't that the main entrance?" he asked.

"Eet is, Meester Wallace."

"Then why is it facing east, you stupid …"

The foreman cut him off as he opened the plans and laid them on the ground. "You asked to change eet, Meester Wallace. Here." The foreman pointed to a change order draft to relocate the entrance of the main house from a southern exposure to an eastern one. It was signed by William Wallace. "See?" the foreman said as he pointed to the signature.

Owen stared at the signature. He turned, pulled his coat tight around his neck, quickly walked back to get in the car and left.

The foreman could see Owen sitting in the back seat looking out the window as he drove by.

"Eediot," he said.

—*—

The hot water was steaming the mirror in the bathroom obscuring Carl's reflection. He reached out and wiped it with a hand towel exposing half of his face with shaving cream and the other half without. His dark beard and skin were a sharp contrast to the white shaving cream. He finished the shave, pulled the head of the razor off and opened the door under the sink to throw it away. He tossed it in and it bounced off the sink drainpipe and fell behind the garbage can. "Great," he said. He leaned down and pulled the small garbage can out to find the blade. He noticed a plastic bag shoved into the corner of the shelf. He reached in, pulled the bag out and opened it. There were several vitamin bottles inside the bag. Carl pulled some out and examined the labels.

Lemon balm—Grape Seed Extract—Theanine Green Tea—Ginsing

Carl read the labels on each. *Improved cognition and memory, GABA transaminase inhibitor properties, increases brain serotonin and dopamine levels ………*

He put the bottles back in the bag and stuffed it under the sink where he found it. He knew it wouldn't do any good to bring this up to Alice. Not now. She became extremely upset when she overheard Carl, Imar and Norman talking about the virus. She cried most of the night. Her fear was real and there was little Carl could do about it. He tried to console her, but the idea of losing her memories was too much for her.

Carl leaned on the counter and hung his head as the water ran and the steam fogged the mirror again. He wished there was something he could do; some way of assuring her she would be alright. But he couldn't. He didn't know if she would be alright or not. This was new territory. It was scary for him, too. *God, help me. I don't know what to do. What if she starts forgetting stuff? What if she leaves the stove on and it catches the house on fire? What if she forgets to ......?*

"What's wrong?" Alice asked as she saw Carl leaning on the counter.

"Oh. Just ... clearing my head some."

"Steam?"

"Yeah. The steam feels good." Carl raised his head and turned the water off. He wiped the rest of the shaving cream off and turned to Alice. She saw how handsome he was with his t-shirt on and a clean shave. His beard was trimmed perfectly. His eyes bright.

"I remember when I first saw you," she said. "You walked into the room at the CDC where we were meeting on 19Q. You stood a foot above everyone else. The place stopped. My heart stopped." She moved closer with the aid of her crutch. "You are a handsome man, Carl Kruger. Indeed," she said as she laid her crutch aside and put her arms around his waist.

Carl put his big arms around her shoulders. "I remember that same moment. You came right up to me and took my arm. Your smile told me you had a tender heart; your eyes captivated me. It was love at first sight, baby."

"It was," Alice agreed.

Carl looked at her closely. "Honey, I want you to know you are the joy of my life."

Alice blushed slightly, then became a bit sullen. "Carl, I'm scared," she said as she buried her head into his chest. She could hear his heart beating. "I don't know what to do."

"Nothing. Don't do anything," he said.

Alice leaned back. "What do you mean?"

"We have two years before any symptoms will surface. Even then, they will be normal progression. The team will probably find a cure by then."

"What if they don't? What then?" Alice was raising her voice slightly, expressing her frustration.

"I don't know, Alice. What I do know is I love you, and as long as I have breath, I will never leave you. We have a faith in God. He brought us together. We have to believe He knows what's going on here."

"But is He going to do anything about it?" she asked.

Carl thought for a second. "I don't know, baby. Sometimes, it's not for us to know."

January 26

CARL WAS EXCITED to see his old friend. He drove up to the three-story building and parked in the large lot in front. He walked past the tall, glass barrier that protected the facility from would-be bombers or terrorists. After the Oklahoma bombing, all federal buildings, especially the Federal Bureau of Investigation, constructed bomb barriers to keep trucks or cars packed with expolsives at a safe distance from their facilities. Three flags flapped gently in the cold, morning breeze.

The entrance funneled into a small area bordered by glass partitions with doors. Two armed guards stood behind the counter. "May I help you?" the older guard asked.

"Carl Kruger here to see special agent Jamille Larson."

The man scanned down a paper with a list of names. "Yes. Second floor. Please empty your pockets into the tray and proceed through the scanner."

Carl followed the instructions and stepped to a scanner where two more armed guards stood. One motioned for him to stop, waited for a green light, and then motioned him through. Carl picked his items off the conveyer belt and headed toward the elevator.

The elevator bell rang as the door slid open. Jamille was standing in front of the door with a big smile. "Well, well, well." He turned to a middle-aged man standing next to him. "I think we better arrest this Nobel Prize winner guy."

"Jamille!" Carl said as he stepped out and hugged his friend. "It's great to see you."

"And you, you big bear." Jamille turned to the other man. "Carl, this is Special Agent Jeremy Harlow."

The agent put his hand out to shake Carl's. "I'm not a hugger," he said with a big smile. Carl laughed as they shook hands.

"Jeremy is the man in charge here and agreed to my temporary assignment."

"Thank you," Carl said.

"Welcome."

"I've told him all about you and your work. He's agreed to provide me with one assistant if I need it."

"Great. I think you will."

"Let's move this into my office," Agent Harlow said as he motioned down the hall. Carl and Jamille chatted about Carl's family and Jamille's latest sort-of girlfriend as the three men walked into Agent Harlow's office. They sat at the table near the window overlooking the bay.

"So, tell me what this is all about, Carl."

After being assured of their confidentiality, Carl proceeded to tell the two men about the development of Cerebtol, Owen's disappearance, how he saw him at the airport in France headed to India and the emails. "Looks like you've done a lot of the investigation for us, Carl," Jamille said.

"Thanks, but I know there's more."

"I don't know if we can get a subpoena to search the files, Carl," Special Agent Harlow said. "there's no crime, yet, and …"

"You won't need a warrant. The Board Chairman has agreed to let you search the records."

"Great!" Jamille replied. "Then we can get started right away."

January 28

THE LOCKSMITH WAS kneeling at the door jiggling the deadbolt with two small picks. He had his baseball cap on backwards and tools sticking out of his back coverall pockets. "Almost got it," he said in a meek voice.

Jamille glanced at the search warrant. It was easy to obtain after Carl filed a missing person's report and they waited the allotted time. Special Agent Jeremy Harlow helped get the warrant to his favorite judge to get it signed. Jamille watched as the locksmith continued to fuss with the lock. "So, do you do other jobs than just the FBI?" he asked.

"Oh yeah. I do whatever I can," said the tiny, middle-aged man. "In this economy, I gotta keep business commin' in or I don't eat. There!" The lock clicked. The man stood to his feet and smiled as he opened the door. "Welcome home, sir," he said with a poor British accent.

"Thanks," Jamille said as he and a young woman walked into the house with their hands on their pistols. The lady turned the lights on as Jamille stood at the entrance. "Looks empty." He glanced up the stairs. "Anyone up there?"

"What if there was, Agent Larson?" the lady asked.

Jamille turned and smiled at her. "Then we would go introduce ourselves." He looked around. "Let's sweep the downstairs first, then upstairs." Jamille and the woman went into different rooms ensuring no one was in the house. They finished the upstairs in a few minutes. "O.K. Empty. Now, we need to see if there is anything here that will tell us when he was last in the house. Look for papers, trash …"

"With all due respect, Agent Larson, I know how to search the premises." The woman had an attitude and Jamille kind of liked that. She was attractive with brown eyes and tanned skin. Her black hair fell to her shoulders.

"O.K. Then you take downstairs and I'll look up here."

The two of them split up. Jamille looked in drawers, closets, trash cans, dirty clothes, clean clothes, medicine cabinet; anything that might give him a clue. He noticed some shirts and slacks were missing from the closet and nothing in the dirty clothes. There was no suitcase or luggage anywhere upstairs.

When he finished he went downstairs to see what Agent Carlson found. "Find anything, Wendy?"

"The trash is empty. There are no bills on the desk, no papers, nothing that I can see."

"Did you look in the refrigerator?"

"Yes, of course I did. Nothing."

Jamille opened the refrigerator. "Don't you believe me? Go ahead and look for yourself then," Wendy said.

Jamille pulled out a carton of half and half that was in the door. "Expires December 23, 2018," he said. "That was six weeks ago."

"Where did you find that?"

"In the door."

Wendy was shocked. "I thought I looked there. I can't believe I missed that."

Jamille chuckled. "Wendy, the only reason I found it is because I missed it once, too. A long time ago. Now, I always look for expiration dates."

"I don't like making mistakes," she said.

"That's how you learn. Get over it." Jamille put the carton back on the door shelf. "So, he's likely been gone for some time. There was hardly anything in the fridge. No vegetables or foods, just older containers." Jamille looked around and saw a calendar on the wall. He walked over and looked at it. "Still on December."

"Are you kidding me? I missed that, too?"

"Take a breath." He walked over to the agent. "Shall we search down here a little more?" he said with a smile.

"Sure," Wendy said dejectedly.

The pair continued to search the downstairs with little success. Jamille found a shredder that was empty, a place where a PED could be charged, a computer tower that was gone and two books taken off the shelf. They searched the garage and didn't find anything more. The car was gone, but they expected that. "I wonder why the car has not turned up somewhere?" Wendy asked.

"Probably at a tow yard somewhere," Jamille said. "Make a note to make some calls."

"Do you think he's gone?"

"He's gone. No doubt about that. He left about a month ago. The expired milk, clothes missing, no trash, the calendar … yeah, he's gone. I just don't know where, yet."

"So, what's next, Agent Larson?"

"Please, call me Jamille."

"O.K., Jamille," Wendy said with a smile. "What's next?"

"Money. We need to follow the money. It always tells the story."

—*—

The work in the lab was long and tedious. Imar Spaan was working with two analysts taking extensive notes as they methodically progressed through the variety of samples lined up on shelves in the storage cooler at the Alzheimer's Disease Research Center. Each sample had to be carefully analyzed and variations mapped.

The three of them were comparing the pre and post inoculation tissue samples of several Phase One patients. He planned to review the tissue samples first, and then progress to blood and cerebrospinal fluids. Only six patients allowed brain biopsies.

"This is going to take months," one of the analysts said. "We've been at this for weeks and have hardly made a dent. I think we need more people," the young woman said.

Imar pulled away from the microscope. "We need absolute control on the samples and tests. If we have too many people, we will undoubtedly lose control to some degree." Imar went back to viewing the specimen through the microscope.

"Doctor Spaan. I think you should see this, sir," the other analyst said. Imar moved closer to the young man and peered over his shoulder. "It looks like the viral cluster in this sample is gone."

"Gone?" Imar asked.

"Yes. There is only one virus in the cell, not a cluster as we have seen before." The analyst moved aside to let Imar peer into the microscope.

Imar adjusted his glasses and peered at the specimen. "Amazing," he said. "There is only one virus."

"Well?" the young man asked.

"It is a single virus. Check the other cells to see if they also have a single virus or clusters, and if they are damaged cells or not."

"You mean the virus may not be functioning like before?"

"Possibly. We need to know if this virus is in a live or damaged cell, when it separated from the cluster and the effect of such," Imar said. "Which sample is this from?" Imar asked.

The analyst looked at the vial. "Phase One, patient twenty-six. It is from the cerebrospinal fluid taken sixteen months after inoculation."

"Sixteen months?"

"Yes."

Imar thought for a moment, then realized how concerned he must have looked. He smiled and patted the young man on the shoulder. "Good. Good job. Make a note, keep searching," Imar said as he turned back to his microscope.

The analyst jotted some notes on a pad and returned to his microscope.

January 31

"YOU ALWAYS SEEM to have the good cases, Carl," Jamille said as he walked into the conference room where Carl was sitting with Imar Spaan. "This guy Pitke is a real winner, I can tell you that." Jamille set a box of documents on the table.

Agent Wendy Carlson walked into the room carrying another box of documents and set them on the floor by the table. She pulled out a stack of papers and handed a report to Imar, Carl and Jamille. "This is our summary report," she said.

"Thanks, Wendy," Jamille said with a smile as she walked out. Carl jabbed Jamille in the side and laughed. "No, Carl. It's not like that," Jamille said childishly.

"Yeah, right," Carl replied.

"O.K., so, back to work here." Jamille opened the report. "We found out Owen Pitke is gone. He left the United States on a trip to six different countries."

"What? Hold it," Carl said. "Six?"

"Yeah. Look here. Owen left December 17th last year. That was when you said you saw him in France headed for India. He went to India alright, and planned to continue on to China, Taiwan, Tokyo, and South Korea."

"An international traveler, huh?" Imar said looking at the report.

"Sort of. He only made it to India from what we can see. The other tickets were never used."

"This is crazy. He wasn't planning a trip abroad, much less to India. Why buy tickets to all of those countries and stop in India?" Carl asked.

"To get away. Hide." Jamille answered. "We have no idea where he is. Our best guess is he is in India or that general area."

"Now just stop. This is nuts," Carl said.

"Just hush for a second and let me tell the story, will 'ya Mister Doctor?" Jamille took a deep breath. "The money always tells the story, Carl. Owen left the country because he was conducting insider trading with Antrole Pharmaceuticals; your competitor."

"Antrole? Did he buy their stock?" Imar asked.

"Yeah, and a lot of it. Four hundred thousand shares over a six month period from June eleventh to December sixteenth, 2015. He made more than fifteen million on the purchase and sales."

"He must have given them the product somehow; stolen it," Carl said.

"I'm sure he did. He had a large deposit a few days before his first purchase by cashier's check from 'Goldfield.' They deal exclusively in gold bullion and coins. I suspect he was paid in gold, cashed it in for a cashier's check, made the deposit and then made the first purchase. He's drained his accounts here to offshore accounts and then likely to Switzerland, we suspect."

"Unbelievable," Carl said.

"No, what is unbelievable is that little Mr. Pitke also participated in a murder for hire."

"Murder!" Imar was shocked. "Murder!?"

"Yes, Imar."

"Who? What?" Carl asked.

"Do you remember when I got involved in the death of Quinton Lemolo?"

"No," Imar said. "No. Not Quinton," he whispered.

Carl and Imar quieted down to hear Jamille. They became sullen.

"Yes, I'm afraid so. We searched Owen's records and found a payment to a company run by a known mobster in Montana; Three-Fingered Jack."

"They have mobsters in Montana?" Imar asked.

"They are everywhere, my friend. The digital age. So, Owen apparently hired Jack to do Lemolo. I thought Lemolo was a hit job, but I assumed it was tied to his gambling. What's puzzling is we haven't found any phone calls to Jack; just the payment through a corporate account."

"Then who contacted him?" Carl asked.

"Don't know, yet. But we will."

"Why?" Imar asked.

"Pitke probably did it to keep him quiet about the medicine."

"What? You mean Owen knew about this two years ago? And didn't tell anyone?" Carl was getting angrier by the minute.

"Yeah. We found deleted emails and voicemail messages from Lemolo. He was panicked about forgetting small things. Pitke probably didn't want anyone to mess up his stock earnings when they went public or made the announcement. Look on page two. He bought the stock at a quarter, sold it off for forty bucks. Look at the dates."

"My Lord," Imar mumbled.

"You mean he had information that could have prevented millions of people from getting the inoculation?" Carl was stunned. His emotions were running rampant. "We could have stopped this. All of this, Imar."

"To think, he did this for money," Imar said solemnly.

"He did," Jamille confirmed.

"Alice," Carl said. "Alice could have avoided getting the shot."

Imar reached over and touched Carl's forearm. "I'm so sorry, my friend."

"There's more, Carl."

"How could there possibly be any more?" Carl asked.

"Don Sinclair also bought Antrole stock. A lot of it."

"Don? From the FDA?" Carl asked.

"Yes, FDA Don," Jamille said. "He made a ton of money on this, so he must have known about Owen and Antrole."

"Did he know about Quinton?" Imar asked.

"Don't know. We're still looking. We only know he bought stock starting the same week Owen did, a lot of it."

"Does he still have it?" Carl asked.

"Yes. And that means he doesn't know Owen is gone, or that there's a problem. He may not know about Lemolo, or knows and doesn't care if he thinks Pitke is still around to take the fall." Jamille said.

"I see here there are quite a few deleted email and messages," Carl said scanning over the papers.

"Yeah. We transcribed the messages and reconstructed the emails. You may want to see them. They're from patients forgetting things and panicking."

Carl laid the papers on the table. "So, Owen knew about problems with Cerebtol all along, starting with Quinton, and hid it."

Jamille cleared his throat. "He did."

"This is horrible. It's a nightmare," Imar said.

"What are we going to do now, Jamille?" Carl asked.

"Well, I'm going to go talk to Mr. FDA Don and see how he reacts." Jamille looked at Carl and Imar. "I suggest you keep researching this drug to see if you really want to keep it on the market. That's your call, not mine."

"Are you still going to look for Owen?" Imar asked.

"Not right now. I expect he will pop his little head up sometime soon. It's just a matter of time. I want to focus on FDA Don for now," Jamille said. "I bet he has some interesting stories to tell."

The Assault

February 3

1:55 A.M.—THE YOUNG, olive skinned man was sitting behind the steering wheel in a car a few blocks away from the Guardian Biopharma Storage facility outside Memphis, Tennessee. He was thin with a scruffy beard and was wearing a green taqiyah[22] with gold and brown embroidered trim. The man had large, brown eyes in a fixed gaze at the building down the street. He checked his watch; 1:56 a.m. He glanced to the dashboard where a small switch was showing. He looked over to the passenger sitting next to him, another olive-skinned man, heavier build and also with a beard. He was wearing a white taqiyah with gold trim. He was much older than the driver. A passer-by could have mistaken them for father and son. The passenger was checking his assault rifle and grenades. Neither man smiled.

The driver turned back to stare at the facility. He closed his eyes and bowed his head. "Allahu Akbar, Allahu Akbar,[23]" he sang quietly in a melody. "Allahu Akbar Fauqua Kaidi L'mutadi.[24]" The passenger joined in as they quietly sang the first verse of the old Egyptian military song and onetime national anthem of Libya.

---

22 Taqiyah—is a short, rounded cap worn by some observant Muslim men. The taqiyah serves as a form of religious uniform to identify oneself to the rest of the world as somebody who adheres to Islam.
23 God is greatest! God is greatest!
24 And God is greatest above plots of the aggressors.

The driver opened his eyes and looked at his watch; 1:57. He looked to the passenger who nodded his head. The young man reached down and started the car. He continued to chant and sing quietly in Arabic as the passenger reached over and flipped the switch on the dashboard. The red light began blinking. He looked behind the seat to see several large containers filled with explosives. A red light on a device placed on top of the containers was glowing brightly. The bomb was armed.

The passenger opened the sunroof of the vehicle and looked at the blinking light. The driver revved the engine. He checked his watch; 1:58. "It is time," he said as he looked at the passenger who again nodded his approval.

Time to die.

The young driver was sweating profusely as he gripped the steering wheel and lowered his head. "Allahu Akbar," he said. The passenger climbed up onto the seat he was sitting in and squatted. The driver looked at the passenger with a look of fear. The passenger pointed a semi-automatic pistol at his head and replied, "Drive."

The driver turned back to look at the road, put the car in gear, and accelerated toward the building. He turned his lights on bright and headed toward the entrance of the facility. A metal arm hung across the road between two, tall metal fences. Two small guard shacks bordered the road at the gate. Each shack had two guards inside. Several monitors hung on shelves above the window. The guards were looking at the monitors observing the perimeter of the grounds when they noticed the car.

One guard in the shack on the incoming lane stepped out and shielded his eyes. He put one hand on his pistol and held his other hand up to signal the car to stop and dim his lights.

The older man stood through the top of the sunroof. "Allahu Akbar," he yelled as he aimed at the guard standing in the road and took one shot. The guard fell backwards.

The driver slid down below the steering wheel and accelerated as two other guards stepped out of their shacks, crouched down and began firing. The windshield in the car exploded as the passenger hurled a grenade toward the guards. The guard in the shack activated a bomb barrier as the grenade exploded, tearing the shack into pieces. The barrier slowly rose out of the ground as the car sped toward the gate. The passenger fired two more shots, killing one guard before he was shot in the forehead. His body slumped out of the sunroof and over the broken windshield, dripping blood onto the driver's face as he sped through the gate. The car jumped as it bounced over the bomb barrier and entered the compound. The guard at the shack ran inside the small building as a helicopter appeared over the trees. A man sitting in the open door of the helicopter fired several rounds from an automatic assault rifle into the shack, killing the guard.

The young driver turned the vehicle across the grass and to the side of the building. He circled wide toward the fence and turned directly toward the building's wall.

"Allahu Akbar," he yelled as he crashed into the wall. The body in the sunroof flew against the wall and splattered blood all over it. The young man sat behind the airbag with blood on his face and hands. He was bleeding profusely from the nose and mouth. He looked at his watch. It was 1:59 a.m. He gripped the steering wheel and leaned his head back onto the seat and looked at his watch as the seconds blinked away. As it changed to 2:00 a.m., the blinking light on the dash went solid, and the car exploded.

The helicopter landed on the grass about fifty yards from the explosion. Three men jumped out of the helicopter and onto the grass. One had an automatic weapon and stood guard while the other two ran toward the building.

The car was gone and in its place was a gaping hole in the side of the building. The two men crawled over the wreckage and through the smoke into the building. The men had automatic pistols in both hands. They climbed over the debris in the room and worked their way to a door on the far side of the wreckage. Water was pouring from the sprinkler system in the ceiling as they ran through the smoke-filled hall to a large room in the middle of the building. They looked inside to see one guard with a panicked look on his face.

One man pulled two small, square boxes from his jacket, pushed them against the top and bottom of a door near the hinges and pressed the buttons. They jumped back and covered their faces. The explosion tore the door off the hinges. They ran into the room where the guard was standing with his hands up. His pistol lay on the ground in front of his feet. The lead man shot him in the head without emotion, stepped over the body and proceeded to the locker.

The men opened the door and quickly examined the cartons. One man pointed to a stack. "Here," he said.

The cartons read;

Cerebtol 2018-04B

"No," the leader said in Arabic. "That is the recall." He looked through the boxes. He pointed to a stack of cartons. "These," he said.

Cerebtol 2017-07A

The men grabbed several cartons and stuffed them into oversized backpacks. They grabbed several more and ran back down the hall, through the entry room, over the remains of the car and to the waiting helicopter. They noticed two more guards lying motionless in the grass. One was face down.

"Hurry. There's more," the lead man said in Arabic. They could hear sirens approaching. The two men ran back into the building and grabbed another load of cartons. They heard the sirens as the police cars got closer. "Hurry." As they neared the helicopter, they saw the lights round the corner and approach the gate.

The three men stood outside the helicopter and watched the cars stop at the gate. The bomb barrier was very effective. The police cars could not pass it. The three men laughed and waved at the policemen as they climbed into the helicopter. The police climbed out of their cars and fired several volleys at the helicopter as it lifted off. They watched helplessly as the helicopter sped toward the horizon and dropped out of sight.

—*—

2:15 p.m.

It was Sunday afternoon and raining. The Alzheimer's Disease Research Center was on high alert. Word came that a raid on the Guardian Biopharma Storage Facility outside of Memphis, Tennessee was hit a few hours ago by a military unit of some type. Seven guards at the facility were killed. One survived with serious wounds. He was alive and conscious and told investigators that the people in the vehicle were yelling in Arabic. The level of sophistication, boldness of the perpetrators and execution of the coordinated plan with the suicide car bomber and helicopter led everyone to believe it was an act of terrorism, possibly executed by Al Tahrir. The military style of the operation suggested such.

It was confirmed that the terrorist group took one item from the storage facility; Cerebtol. They took enough to provide fifty thousand doses.

The theft put everyone on edge, especially Carl. He was with Baltimore Police Lieutenant Meyers, a wiry man with glasses and a mustache, and Sergeant Willows, a gruff looking man in his mid-fifties, discussing the raid on Guardian Biopharma. They agreed it likely occurred because of the FDA recall announcement all over the news and the Internet, and the thieves targeted just the Cerebtol. They were discussing adding permanent security measures to the research center since their product was the lone target of the raid. The policemen suggested adding extra security outside, installing bomb barriers, a fence, and possibly asking the National Guard to temporarily provide some assistance. They suggested Johns Hopkins take immediate actions because of the very real threat while developing a long term plan. Carl knew the Johns Hopkins board would have to authorize the additional costs of any security measures he recommended, including relocating the entire operation. Now he had this to deal with on top of the viral research.

He could tell it was going to be a bad day.

"Carl," Jamille said as he approached the men. "I'm very sorry to interrupt."

"Gentlemen," Carl said. "This is Special Agent Jamille Larson from the FBI." The men greeted each other with handshakes.

"Special Agent?" the lieutenant asked. "What is the FBI doing here?"

"Jamille is a personal friend of mine."

"Oh," the lieutenant said. "I thought he was here because of the raid on Guardian Biopharma."

"No. But, I am glad I'm around so I can offer some advice to my friend," Jamille replied.

"And what advice would that be, agent?" the Sergeant asked.

"Retire," he said with a laugh. The others joined in.

Carl noticed Jamille had a look that he wanted to speak to him in private. "Well, gentlemen. Looks like we've beat this horse to death."

"We have, Doctor Kruger," the lieutenant said. "We'll set up some regular patrols and check in with you later this week."

"Thank you, lieutenant," Carl replied. He watched as the two men left the area. "What is it, Jamille?"

"We just arrested Don Sinclair for securities fraud and conspiracy to commit murder."

"Murder? Don?"

"Yeah. We tied him to Lemolo's murder with Pitke."

"Unbelievable," Carl exclaimed.

"He confessed everything this morning; even Pitke's involvement."

Carl was right. It was going to be bad day.

—*—

Imar called the meeting. He wasn't in the habit of exaggerating issues. He was precise, controlled, a scientist. The only people in the meeting were the three Nobel Prize laureates, Imar, Carl and Norman, and Jamille. Jamille had become an honorary member of the inner circle of scientists. He revealed the conspiracy of Owen Pitke and Don Sinclair to cover up the damaging emails of patients experiencing significant memory deterioration. He tracked the financial transactions to show they both conspired to buy hundreds of thousands of shares of Antrole Pharmaceuticals at pennies and sell them for dollars, making millions. He knew as much or more than the scientists about the situation. He was also their trusted friend. He worked closely with them on the 19Q disaster years ago. They valued his opinion and greatly appreciated his work.

"I'm not sure how to start this," Imar said.

"We know it has to be bad news, Imar," Norman said.

"Just say it," Carl encouraged him.

"The virus has changed … significantly," Imar said.

"How so?" Carl asked.

"We have verified the virus is no longer existing as a cluster group in the damaged cells."

"We knew that," Carl said.

"Yes, but what we discovered is that it is able to exist as a single virus … in a healthy neuron cell." No one said a word. "It has mutated, again."

"I don't understand," Jamille said. "There's a new virus?"

"Yes," Imar said.

"A live cell?" Norman asked.

"Yes; a live neuron cell."

"Do we know how?" Carl asked.

"Likely from the chemical element Maskill created," Norman said. "We've verified it contains the element, so I suspect it has caused some type of delayed mutation, similar to what we saw in the wheat with 19Q."

"I'll tell you, Carl, if it's not a mutated wheat chromosome, it's a mutated virus. You really know how to pick them," Jamille said.

"Not funny, Jamille," Carl said.

"Do we know when it mutated, Imar?" Norman asked, ignoring Carl and Jamille.

"Sometime while in the body," Imar responded. "We narrowed the timeframes of the samples and our best estimate is a year or so after inoculation."

"Alice received her inoculation about a year ago," Carl said. "I need to call her." He started to pull his cell phone out when Imar reached over and touched his arm. "Wait, Carl," he said. "Wait a second. I think we need to discuss how this new virus changes everything."

"What? What do you mean?" Carl asked.

"The virus is not killing the healthy cells; it is only existing," Imar said. "If the virus has mutated in Alice," Imar explained, "then it has likely infected other healthy cells by infusing itself to the cells. That means …"

"If we kill the virus, we kill the healthy cells," Carl continued.

"Correct," Imar said.

"And the state of the patient is worse," Norman added.

Imar and Norman stared at each other for a few seconds. "Correct," Imar said.

"Dear Lord," Carl said as he stood and walked away from the table. "What have we done?"

"We didn't push for the approval and distribution of Cerebtol, Carl," Imar reminded him.

"But we discovered it. We applied it to cure a disease. Imar, we received the Nobel because of it!"

"I am well aware of that, Carl."

"O.K., hold it," Jamille said raising his hands. "It's no use blaming anyone for this. If you want to blame someone, blame FDA Don or that pitiful Pitke. They pushed this for their own benefit."

"We know," Carl said dejectedly. "We were fools."

"If we kill the virus in the recipients who had significant cell damage at the time of inoculation," Norman said, always being the scientist and ignoring their discussion about blame, "then they revert to their prior state."

"Unless the virus mutated and expanded into living cells," Imar added.

"Then they will be in a worse state," Carl said.

"If we kill the cell in a recipient who had no damaged cells and took the inoculation as a preventative measure, then the damage to the brain from killing the virus is negligible," Norman said.

"Unless the virus has mutated and spread," Imar said again, "in which case the recipient could have a severely infected brain. A single virus is capable of infusing into a healthy cell. If it migrates to other cells …"

"So, we need to see if the virus has mutated in the recipient and spread before we decide to kill it or not," Carl said.

"We would have to determine if the virus is still existing in clusters in the recipient and the level of infection before we could determine if the condition of the patient would be better served by killing the virus or not," Imar said.

"So, if I am hearing this right," Jamille said, "if the virus has mutated, either way the patient is toast." The men looked at Jamille as he continued. "Unless it hasn't spread," he added.

"I think that sums it up, Jamille," Imar said.

"We need to find a way to kill it," Carl said.

"That shouldn't be too difficult, Carl, since it is similar to the original JC virus that our body so effectively combats."

"Yes, but we've made it harder to kill with the protein shell," Carl said.

"We can develop an enzyme inhibitor that will allow the person's anti-bodies to naturally kill the virus," Norman said. "It's the basics of medicine."

"It is," Imar agreed.

"I feel like I'm in the presence of brilliance," Jamille said. "Did you guys know you are speaking in a foreign language?"

The men ignored his comment.

"Then we need to develop the inhibitor right away," Norman said.

"And a program to analyze the mutation existence and infection levels of the patients," Imar said. "That will likely be a biopsy."

"Of the brain?" Jamille asked. He was shocked.

"Yes, Jamille. Of the brain," Carl said. "It's a Stereotactic Brain Biopsy. It's almost painless."

"I don't want to know."

"We can't expect to do a biopsy on every patient to see if we should kill the virus or not?" Norman said. "That is unrealistic."

"I know!" Carl yelled.

"Easy there, Carl," Jamille said.

Carl walked away from the group, then turned back. "Listen, this is a disaster. We have to start somewhere. The only way we can determine if the patient is able to receive something to kill the virus is to verify the virus still exists as a cluster and hasn't spread. That's through a biopsy. Do you have any other suggestions?" Carl asked as he glared at the trio.

No one said a word. They could see Carl's frustration and near anger, and rightly so. There was no other simple alternative. "No, Carl," Imar replied. "Unfortunately, you are correct."

"O.K. Then it looks like we have a plan to at least address a few patients and see if this works or not," Carl confirmed as he walked back to the table. "I need to let the chairman know what we are doing and pull Cerebtol off the market."

"Agreed," Imar and Norman said in unison.

"I need to tell Alice, too."

"I wouldn't do that Carl. Not right now," Imar said.

"Why not?"

"I think it would be better to wait until we know how to effectively stop the virus. We could have her take some tests, maybe a biopsy if she consents, so we can determine the," Imar paused, "level of infection before we determine the appropriate action."

"The longer we wait, the virus has a better chance of mutating and spreading," Carl said. "What if it spreads to other brain cells and not just neurons?" The room became eerily quiet as they pondered the catastrophe that would cause. People would not only lose memory, they would lose all brain functions, and die. "Are we willing to take that risk?"

"If that occurs, then we are talking about much more than just Alice, and much worse than just memory loss, Carl," Imar said. "There is nothing we can do right now to stop it. We don't have a way to kill it, so why worry Alice about the possible result or a future mutation without knowing how to stop it?"

Carl thought for a moment. "Jamille, what do you think?"

"I agree with Imar. When you know how to approach this thing, then you can tell her. That would give her hope instead of fear."

"One month," Carl said. "No more."

February 9

THE MEN WERE sitting in a circle around a small campfire in front of a large cave. They were wearing long galabiyyas[25] over white tunics and pants. Each man had a long beard and a turban. Several younger men wearing coats, pants and hats were standing guard at various points around the encampment with assault rifles and radios. The older men in the circle had their hands cupped in front of their mouths in prayer. The elder of the group finished his prayer and the rest followed his lead. He spoke softly and clearly in Arabic.

"We have found favor with God, brothers." The men smiled and nodded in agreement. "The Great Satan will fall to us, and we shall wipe the infidels from the face of the earth. God has told me that we are to be patient as a lion, stalking our prey and lying in wait for the right time. Our weapon shall be our enemy's ignorance and our intelligence." The man motioned to one of the guards who carried a small, metal box and set it in front of the leader.

"This, my brothers, is our future." He opened the box and a cloud of cold air spilled out onto the ground as the warm air hit the dry ice. The leader lifted a canister out of the box and held it up. "Our enemy has created the weapon with which we shall destroy them." He motioned for another man who came to his side and sat down. He had a small pouch that he placed on a pillow in front of the leader. He opened the pouch and pulled out a syringe.

---

25 Galabiyya (Jalabiya)—Long outer robe, neck to ankle, worn by Muslim men.

"Our enemy has created a medicine that increases a person's intelligence and memory. They use this to cure their elderly and feeble. We shall use this to out-think our enemy and devise ways for their destruction that they never imagined." He held up a small vial that he removed from the canister. The firelight reflected off the glass, making it look similar to a jewel. The men smiled as they gazed upon the small vial. A tiny label was on the bottom of the vial.

Cerebtol-2017-07A

The leader handed the vial to the man sitting next to him holding the syringe. The man took the vial, inserted the syringe into it, and withdrew a small amount of fluid. The leader held out his arm for the man seated by him who tied a rubber strap around the arm, found a vein and slowly inserted the needle.

"I will be the first to submit myself to the will of God," the leader said as he looked at the group, never taking his eyes off them. The man pulled back the plunger to make sure he hit a vein.

"God is great," one man said as he watched the deep, red blood mixed with the clear fluid in the syringe.

"I submit myself to the will of God and to jihad," the leader said, raising his fist as the man pushed the plunger, injecting the leader with Cerebtol.

The men in the group clapped and praised God for their leader, a brave man who was willing to be the first, the example for their people.

"Every man, woman and child who believes God has ordained us to purge the earth of the Great Satan shall follow my example and receive the medicine that shall make us a great people. Praise God!"

The men in the group shouted with joy and started to roll up their sleeves to receive their injections.

Jihad had begun in front of a cave in the mountains bordering Afghanistan and Pakistan.

February 13

MARGARET GIDDINGS WAS not an easy sell. Carl told her about the mutated virus, how it was now in living cells and capable of spreading. He told her about Owen Pitke and his knowledge of the patients' memories deteriorating over time. He told her about Don Sinclair and his insider trading, along with Pitke. He told her everything he knew, yet she was still hesitant to pull Cerebtol off the market. Carl believed she may have also known about these issues and wanted to separate herself from Owen and Don, or maybe she was buying time. For whatever reasons, she was wrong.

"Ms. Giddings. With all due respect, we have to pull this product."

Margaret stroked the Calico cat sitting on her lap. The cat purred so loudly Carl could hear it though he sat across the room from her in the guest chair. "Doctor Kruger. Pulling that product will have dire consequences for the University and the Foundation."

"I understand, but to leave it on the market is unconscionable," he said. "It is not helping anyone and will only make things worse."

"I agree," Imar said. He was sitting between Carl and Margaret creating a triangle in the room. "This virus is changing as we speak. We have already confirmed changes in Phase One recipients and are analyzing Phase Two."

"And how long will that take, Doctor?"

Imar looked at Carl. "I expect we will have answers in … two weeks?"

"Yes, two maybe three," Carl confirmed.

"Then I don't see a problem with delaying just a bit more before we announce a complete withdrawal of the product," she said as she smiled at the cat.

"Is there a reason for the additional delay?" Carl asked.

"Yes, there is," she said as she stroked the cat. Carl and Imar waited to hear what it was, but Margaret never volunteered the information.

"And?" Imar asked.

"And what, Doctor?"

"And the reason is?" Imar continued.

"What reason?" she asked.

Carl looked at Imar with puzzlement. "The reason for the delay, Madam Chairman," Carl said.

Margaret stopped stroking the cat and looked directly at Carl. "What delay, Doctor?"

"The delay of withdrawing Cerebtol," Carl said.

"Oh. Oh, yes. Cerebtol." She started stroking the cat again. "Oh, just because is the best reason I can give right now." Margaret stood and the cat jumped to the floor. "Now, gentlemen. If you will excuse me. I have another appointment that I must get to."

Imar and Carl looked at each other as Margaret started toward the door. "Uh, well, I guess we will continue this conversation at another time, then," Carl said as he placed his teacup on the table.

"Yes, quite," Margaret confirmed. "Another time," she said as she opened the door, waiting for them to leave.

As the men were walking out the door, Imar stopped and turned to Margaret and extended his hand. "Thank you, Margaret. Would it be alright if we met tomorrow, Saturday to finish this conversation. Say about two?"

"Yes, that would be fine," she replied.

Carl and Imar stepped outside as she closed the door.

"Tomorrow is Thursday," Carl said.

"Yes, I know," Imar responded.

"I can't wait, Imar. I need to contact the FDA and inform them of the situation."

"You may get fired."

"My job is the least of my worries right now."

February 22

FDA NEWS RELEASE

For Immediate Release: February 22, 2019
Media Inquiries: Jennifer Sands, 301-555-4666,
jennifer.sands@fda.hhs.gov
Consumer Inquiries: 888-INFO-FDA

The United States Food and Drug Administration is issuing a mandatory recall of Cerebtol, a vaccine used to prevent dementia in people with either Alzheimer's Disease or Early Onset Alzheimer's Disease manufactured and distributed by Timore Pharmaceuticals, Inc.

The U.S. Food and Drug Administration directs anyone in possession of the vaccine to please contact Guardian Biopharma Storage Services at 888-555-2233 for return instructions.

THIS IS A MANDATORY RECALL.

Cerebtol is marketed in the United States by Timore Pharmaceuticals, Inc., a recipient of and partner with the Howard Foundation of Johns Hopkins University.

On June 20, 2017, the FDA approved Accelerated Development Review of Cerebtol to reduce and reverse the effects of Alzheimer's Disease. **The Accelerated Development Review of Cerebtol is currently suspended pending inquiry.**

—*—

Carl's phone was practically ringing off the hook. He placed the phone on "do not disturb" and informed his secretary to take all messages. He already received a call from Margaret Giddings who threatened to fire him. He didn't care. Besides, she couldn't terminate his employment. Only the dean could do that and he was in support of Carl's decision. She threatened to have the board remove the dean. That wasn't Carl's issue. He was more concerned about the research results of the samples taken from the Phase Two patients. Imar and Norman had just walked into his office when he hung up the phone. Carl looked up to see the two men. It was obvious they were disappointed.

"It's changing, Carl," Imar said.

"Again?"

"Yes," Norman said. "It's actively spreading in Phase Two recipients."

Carl leaned forward. "Actively? What do you mean 'actively'?"

"We have confirmed the virus now exists in a live neuron cell and is spreading throughout the brain as a normal virus would," Imar said. "It multiplies in the cell, kills the cell by exploding it, and infects the surrounding cells. That is why we are seeing accelerated deteriorating memories in the recipients."

Carl looked at the two men. He was speechless.

"Carl, the viruses that we injected to grow in the damaged cells have taken hold and are now capable of killing the good cells," Imar said, "… at alarming rates, Carl."

"We've created a nightmare," Carl said staring at the desktop. "Do we know why?"

"No. Maybe it's some gamma rays from a solar flare, or interactions with another chemical that's affecting the virus. All we know is it's happening," Norman said.

"Can we stop it?"

"We think we can, pretty easily, but it will likely kill the good cells that it inhabits because the virus has infused itself into the electrical impulses of the cell," Norman said. "If it hasn't multiplied enough to outright kill the cell, then it has infused itself into the cell."

Carl walked over to the window. He loved the view of the bay. It helped to clear his head to think. "That's it?"

"We have an enzyme inhibitor that appears promising. The success rate with the rats and ferrets is high." Norman said.

"We need solid data to get the FDA to approve testing on our groups." Carl said. "If we let it go, it kills the cells through infection. If we kill it, it kills the cells it has currently infected, both good and bad. Either way, the patient decompensates." Carl looked back at the men. "Pretty poor options and no solution from what I see."

"There is really only one option, Carl," Imar said. "We should try and slow the spread of the virus with the inhibitor until we find a definitive way to kill it without damaging the other cells, if possible."

"Then we need to get the data to the FDA right away to use our test group," Carl said.

—*—

Alice was washing a few pots and pans after dinner. Joey was off to his friend's house for a sleepover. Carl was trying to build up enough courage to tell Alice. This was going to be devastating. She was under a lot of stress just trying to cope with her physical limitations and Joey. She was getting older and her mobility was diminishing. She was losing muscle strength and had to work hard at even the simple tasks; walking, standing, washing dishes and the like.

And Joey. He was ten years old soon to be eleven and becoming a handful. Alice would tell of their frequent arguments and Joey's defiance. He was asking about his 'real' mom and dad. He wanted to know who they were, where they were from, where he was from. Alice and Carl told him they were his mom and dad because they were raising him and teaching him their values and about God and education and right and wrong. But he struggled with his identity. It was hard for Alice and Carl. The long hours Carl had to put in. The missed games, events and weekends. Joey was beginning to resent it. It wasn't easy for anyone.

And now Carl had to tell Alice about Cerebtol.

Alice laid the dishtowel on the counter and grabbed her crutch. She waddled her way into the living room and plopped onto the sofa. Carl watched as she put her feet up and let out a sigh. He cautiously moved over to the sofa and sat next to her, rubbing her feet. "How are you doing, baby?" he asked.

"Oh, I'm O.K., I guess." Alice smiled as Carl rubbed her feet. "Getting better every minute."

"Good." Carl rubbed her feet slowly. "I'm glad Joey has a sleepover tonight. You need a break."

"Ummmm."

"We need a break. Things are just crazy."

"Yeah." Alice was relaxing and enjoying the foot rub.

Carl pulled her socks off and massaged her toes and the bottoms of her feet as he talked. "The FDA pulled Cerebtol off the market today."

"What? They what?"

"They pulled it," Carl said.

Alice pulled her foot out of Carl's hand. "You didn't say anything was going wrong with it."

"I know. I didn't want to scare you."

"Scare me?" Alice said sternly. "You don't think this is scaring me?"

"Alice, I …"

"Why did they pull it? What's wrong?" she asked as she sat up.

Carl looked at her. He could see the fear and anger in her face. "You are the best thing in my life. You know that, don't you?"

"Carl, don't change the subject. Of course I know that. What I don't know is what's going on."

"I don't know if …"

"You can tell me. I'm a big girl, Carl," she said boldly.

Carl waited, unsure if he could tell her, but he had to. "The virus has changed."

"Changed? How?"

"It's mutated somehow. The clusters have separated and the virus exists as a single virus in good cells … and starting to multiply like a normal virus."

"Killing the cells?"

Carl turned away and looked at the coffee table. He looked back at Alice. "Yes. It is killing the neurons."

"No!" Alice said as she wrapped her arms around Carl. She squeezed him tightly for what seemed like minutes.

Carl put his arms around her and whispered, "It's O.K."

Alice pulled away. "What do you mean, 'It's O.K.'? It's not O.K., Carl. How long have you known this?"

"I just found out today from Imar and Norm."

"But you knew there was something happening awhile back, didn't you?"

Carl waited. He had to tell her. "Yes, but it wasn't confirmed."

"Confirmed? Carl, you make me so mad sometimes." Alice grabbed her crutches and started to get up. Carl grabbed her arm and pulled her back toward him. "Let go!" Carl held onto her arm. "Let me go!" Alice yelled as she struggled in vain to pull herself free. She started to hit Carl on his big arms. "Let go of me. Let go ..." She stopped struggling as Carl pulled her into his arms. "Let go of ..." she whimpered. "Hold me. Hold me, Carl," she sobbed.

Carl wrapped his big arms around her and started rocking her gently. "I'm scared, Carl," Alice said as she quietly cried.

"Me, too."

—*—

A cloud of dust could be seen traveling along the ridgeline toward the village. Two men on a rooftop near the entrance to the village stood watch toward the mountains watching the old pickup approach. One man peered through binoculars as the other watched the cloud getting closer. As the truck came into view, the man with binoculars could see three men sitting in the cab of the truck scrunched together. He also saw two men standing in the back of the truck peering over the cab holding assault rifles. They were wearing turbans on their heads, had short beards and big smiles.

"Amir!" the man said as he lowered his binoculars. He spoke rapidly in Arabic. "It is my cousin, Amir!" he shouted to the men standing in the street below. They, too, were well armed.

One man was squatting and talking with a little boy, maybe ten years old. He was showing the boy his automatic rifle when he heard the man on the roof yell. He smiled and motioned to the group to clear the streets and for the little boy to go back to his mother who was standing off to the side wearing a burka[26] and holding a baby girl. The woman ushered the little boy into the house and closed the door.

The pickup slowed significantly as it rounded the corner and approached the village. It stopped in front of the group of men. One man walked up to the driver side and spoke to the driver. He looked back at the group with a big smile and gave a "thumbs up" sign. The man who was talking with the boy, the leader, raised his arms and yelled, "God is Great!" He waved the truck on to a square in the center of the village under a large tree. Several people were seated near the tree and standing outside the buildings as the pickup approached. The truck stopped under the tree.

The men jumped out of the back of the truck as the cab emptied. People started spilling out of the surrounding buildings and approached the truck. The leader climbed into the back and stood facing the growing crowd. Behind him were three large Coleman coolers and a couple of boxes. "Today, our God has delivered us from the infidels!" he shouted. The crowd cheered, clapped, and raised their arms. "In these boxes is a gift from God that will give us the power to remove the infidels from this earth." Everyone clapped and yelled again. "We are all to partake of this together, man, woman and child, and become one people like none other on earth, thanks be to God!"

He motioned to a man to open the boxes. Inside were packages of syringes. The leader opened a cooler and lifted a small box of vials. He handed them to the man who removed one vial and inserted a syringe into it, withdrawing the clear fluid and filling the syringe. The leader rolled up his sleeve and faced the crowd.

"In the name of God!" he exclaimed, as the man plunged the syringe into his arm and injected the liquid. The leader raised his arms and yelled, "God is great! Praise be to God who defeats the infidels!" He fired his rifle

---

26 Burka—Also spelled Burqa—a full body cloak worn by some Muslim women.

into the air to the applause of the crowd. He jumped down from the back of the truck and motioned for people to line up.

Men rolled up their sleeves; old, young, children. Everyone wanted their shot. Three women approached the truck and removed a box of needles and a cooler and went into a building followed by a line of women waiting for their injection.

When everyone received their injection, the three men climbed back into the cab and the two men back into the bed of the truck. The truck sped off in a cloud of dust headed for the next village.

Construction

March 4

THE TRUCKS WERE lined up on the driveway waiting for the guard to let them through. Owen stood on the balcony overlooking the entryway to the house and grounds. He could see over the treetops to the road near the bottom of the hill nearly a half mile away. There were two more trucks driving toward his new house. Owen sipped his coffee as he watched two guards climb into the back of the truck and disappear from sight for a few minutes. They reappeared and motioned for the main guard to the let the truck through. The gate opened and, with a billow of smoke, the truck drove up the last leg of the driveway and circled around to the front door. A man standing at the doorway watched as the driver and passenger climbed out of the truck and started unloading it.

Faiz stepped outside and onto the porch. "That goes in the living room, west wall," he said fluently in Dzongkha. The workers carried the teak hutch through the doorway and into the living room. Next came the table with highback chairs, the television, sofa, chairs, beds, and all of the necessities Owen Pitke would need to live a life of luxury. Faiz directed each worker, each truck, every piece of furniture to its proper place. He made a map of each room showing exactly where each piece went to ensure there was no mistake. He was tired of Owen accusing him of making mistakes when, in reality, it was Owen making the mistakes.

"Yes, over there," he said as the workers carried a large desk through the door.

Owen strolled down the stairs as the men carried the piece through the doorway and into the side room. "No, not there," he said as he stopped midway down. "Feliz, that is not the office."

"Faiz, Meester Wallace, and, yes, you told me to make thees room your office, sir." Faiz held up the paper so Owen could see he was using a guide.

"You have the wrong one," Owen said as he met Faiz at the bottom of the stairs. He grabbed the paper from his hands. "See, this is the ..." He stopped and looked at the paper, then looked at the foyer and rooms to each side. He looked back at the paper and turned slightly, orienting himself to the room where the men stood with the desk in the doorway. "Damn. It is that room. I could have sworn ..."

"Eeet is okay, Meester Wallace," Faiz said as he slowly took the paper out of Owen's hands as if it was contaminated. "We will have your fine, fine house all decorated and ready to go very soon, sir." Faiz ignored Owen as he turned to the workers. "Go ahead," he said. "Far wall."

Faiz watched as Owen turned and walked slowly back upstairs sipping his coffee. He became engrossed in the shape and color of the mug as he walked. Faiz shook his head, and studied his paper as he walked into the next room.

—*—

It was eleven at night. The analyst walked up to the desk where Carl had laid his head on his arms for a second, falling fast asleep. It was a long day. It started with driving Jamille to the airport at five in the morning to catch his flight. He enjoyed having his friend here. He did a great job of gathering the evidence to prosecute Don Sinclair and reveal the coverup he and Owen conducted. The information was invaluable to the research.

Carl had been working nearly day and night searching for a way to stop the spread of the virus without killing it. The task was insurmountable, but they had to find a way. If not, the virus would continue to spread, killing neuron cells and leaving millions of people robbed of their memories, or worse. Alice would be one of the victims.

"Doctor Kruger," the young lady said as she lightly shook his shoulder. No response. "Doctor Kruger?"

"What?" Carl sat straight up and looked ahead. "What?" He rubbed his eyes. "Oh, man. What time is it?" he asked as he strained to see the clock.

"Eleven. Most everyone has left."

"Eleven?" Carl put his glasses back on. "Is Doctor Spaan here?"

"Yes. He asked me to see if you could meet with him in lab C right away. You weren't answering the phone."

"Yeah," he said sheepishly. "I was busy."

"I saw," the girl said with a chuckle.

"Lab C?"

"Yes."

Carl slowly got to his feet stretching his muscles. He was stiff and sore from sitting at the microscope most of the day; in fact, most of the month. "Alright. Tell him I'll be right there."

—*—

Imar was sitting next to a centrifuge reading some printouts when Carl walked in. "You called?"

"Yes. We've stopped it," Imar said.

"Stopped it or killed it?"

"Stopped it," Imar said as he handed Carl a printout. "We know we can kill it. Of course, to be technical, a virus isn't really alive so we can't …"

"Imar! I don't need a lesson on viruses," Carl said sternly.

"Yes. Uh, we prefer to not 'kill' it yet. We need to know how to do that without killing the cell it has infused." Imar pointed to the paper. "But we've stopped it from progressing. This is the population level of the virus in an infected ferret. We've been experimenting for weeks on ways to attack the virus by weakening the shell, causing it to die naturally without the use of chemicals and without killing the host cell. Most of our research has been centered on enzymes, proteins, and amino acids." Carl glanced over the report that Imar handed him as Imar handed him a second report and continued. "These are populations today in the same animal."

Carl studied the reports. "This shows neutral growth rates."

"Exactly," Imar said. "The enzyme inhibitor we applied has proven to be successful to stop the spread of the virus, at least in ferrets. I think we have the data we need to go to the FDA and request application to the test groups."

"But won't the virus continue to kill the cells it has infected?"

"No. The virus will just exist inside the cell, functioning as it has before, without killing the cell."

"Are you certain?"

"No."

"No? What do you mean, 'no'?" Carl asked.

"This is a hypothesis, Carl. I have evidence to believe this is what is happening, but I will not know until we test it on living subjects and draw samples of cells to compare, just like on the initial test groups."

Carl tossed the papers on the table in disgust. "Tests? More tests?" He paced across the room and kicked a chair over. "How many more tests? When are we going to stop this thing? What if it mutates into some other disaster? Then what?"

"Carl. I am doing the best I can!"

Carl turned and started toward Imar, angry. His eyes were fixed as he yelled, "Your best isn't good enough! People can't wait, Imar! Alice can't wait!" Imar was shocked at the display and started to scoot away in the rolling chair. Carl saw the fear on Imar's face and caught himself. "Imar. Imar, I'm sorry. I didn't ..."

"It's O.K., Carl," Imar said as he stood. "I understand the pressure you have with all of this; with Alice." He placed his hand on Carl's big shoulder and looked him in the eye. "Carl, I want to find a way to stop this just as much as you or any man. We have something here. We need to get the FDA to allow application to the test group for definitive testing. You know that."

Carl lowered his head. "Yes, I do." He walked back over to the desk, picked the papers off the floor and reviewed them. "If these figures are accurate, then we should see results in a short time."

"Agreed. Let's have the FDA allow us to apply the inhibitor to a few patients in the Phase One group. They seem to be deteriorating the most."

"They should level off and stabilize," Carl said.

"Yes, they should."

"If we can stop its spread, we have time to deal with it, maybe develop a way to make it effective again. At least we have some time," Carl said almost dejectedly.

"We do, Carl. How much, I don't know. Maybe years; maybe months."

"O.K. I'll start making calls. Every day we delay the virus will spread. If we can stop it and cause it to lay dormant, then we should do so right away."

"You are sounding a little like my old friend, the scientist, the doctor."

Carl looked at Imar with great sadness. "I don't feel like him."

March 28

THE MERCEDES PULLED up in front of the curb at the market in Thimphu, Bhutan. The driver opened the door for Owen to step out. Faiz stepped out the opposite side. "They have excellent food here, Meester Wallace," Faiz said as he motioned toward the stalls lining the sides of the alley. "You will find everyting you want here to satisfy your hunger," Faiz said as he rubbed his stomach. "I will get some supplies while you have lunch."

"Sure," Owen said as he walked among the vendors sitting on the ground or on a stool trying to sell their materials, crafts, foods and meats. Faiz watched as Owen ventured farther into the maze of vendors and out of sight. Faiz motioned to the driver to wait as he walked into a bar across the street from the market.

Inside, the room was filled with a haze of cigar smoke. An old juke box played a slow song. Across the room a pudgy man with dark skin, mustache, gray hair and a large beer sat at a table in the corner. His eyes made him look like he was of Chinese descent. Two men stood at each doorway. The men stopped Faiz as he entered the bar. The man at the table motioned for the men to allow Faiz in.

Faiz greeted the man in Dzongkha. "Greetings, Shamar." Faiz bowed to the man who raised his hand to stop him.

"English. I need practice," the older man said.

"Greetings, Shamar," Faiz said again.

"Is the American here?"

"Yes. He is feeding his stomach at the market," Faiz said as he started to sit at the table. The man raised his hand to stop him.

"I sit alone," he said. "Is he with others?"

"No. He is alone."

"Does he have people who call?"

"Those would be friends, Shamar. No. No one calls him. He is alone."
Faiz watched as the man lifted the beer and took a large gulp.

"He is rich. He has 'friends' somewhere." A waiter brought a dish of
noodles with some type of meat and vegetables in it. Shamar began stirring
the contents with chopsticks as he spoke. "I do not like other people here
who are rich. Only me."

"I understand, Shamar," Faiz said.

"Does he sell opium?"

"No, no." Faiz chuckled. "No, he does not."

Shamar picked up a wad of noodles and took a big bite. He had some
noodles hanging out of his mouth as he gulped them down and drank some
beer. He wiped his mouth, leaving a little food on his mustache. "Where he
get his money?"

"I don't know. He just has it," Faiz said.

Shamar pushed the food aside. "When he come back to market?"

Faiz cleared his throat. "Whenever you like, Shamar. I can bring him
here anytime."

"I think about it," he said as he drank the last of his beer and let out a
burp. He pulled out a gold coin and laid it on the table. "Go now."

Faiz picked up the coin and smiled. "Thank you, Shamar," he said as
he walked out of the bar and across the street to the car. He waited for
Owen to return from his lunch. "Good food, Meester Wallace?"

"Sure. Just peachy," Owen said as Faiz opened the door and Owen
climbed into the car, followed by Faiz on the opposite side.

Shamar watched from the doorway of the bar as the Mercedes sped
away.

—*—

Alice sat in the chair wearing the headring for the Stereotactic Brain Biopsy.
She knew how the procedure worked and was not afraid. Carl sat next to
her holding her hand as the doctor dialed in the image and injection point

for the biopsy. The monitor showed a small cross in the center of a triangle. "This is the area we will be extracting the sample from, Alice," she said as she pointed to the monitor. The woman was pleasant and had a slight English accent. She was middle aged and looked professional with her auburn hair framing her round face. Her smile was infectious. "You will feel a sting as I numb the area, then some pressure as I drill a small hole for the biopsy."

Alice sat still with the headring on. She tried to focus on the doctor who was sitting just out of sight to her right. "I know, doctor. Let's get this over with."

The assistant finished cutting away a small patch of hair from Alice's head and handed the doctor the syringe. The doctor inserted the needle and pushed the plunger. "Umph," Alice said as the anesthetic stung.

"Sorry. It will numb pretty quickly." The doctor waited a few seconds and poked Alice's head. "Feel anything?"

"No, just your tapping on my head. No pain."

"O.K., then. Here we go." The doctor started the drill. Alice could hear the high pitched whine. She looked over to Carl who smiled and held her hand. Alice squeezed his hand as the drill started its penetrating journey through the skull.

—*—

Alice was resting comfortably on the sofa. She had a small bandage on the spot where the doctor entered the skull to take the biopsy. Carl knew he had to watch Alice for nausea or vomiting and to make sure she was periodically awakened and coherent with no headaches.

Joey was upstairs with one of his friends playing video games. He was at the age where he could easily entertain himself, especially if he had a friend and a video game. Carl was in the kitchen when the house phone rang.

"Hello?"

"Is this Doctor Kruger?" asked a man's voice?

"Yes, it is. And who is this?"

"I'm the guy that's gonna get even with you for destroying my life," the man said. Carl could tell he was elderly.

"What? Who is this?"

The phone went dead. Carl scrolled through the numbers to see who called. The number showed as 'blocked.' Carl put the phone down and thought about calling the police.

"Who was that?" Alice asked.

"Wrong number," Carl said. He walked into the living room and sat next to Alice, and turned the TV on.

Justice

April 2

ALICE SAT IN the chair looking at the pictures on the wall. She had been in Carl's office many times before, but she never noticed the decorations. She liked the black and white prints of Yosemite Falls and Half Dome. She knew Ansel Adams was one of Carl's favorite photographers. He was more than a photographer; he was an artist. She remembered one time when Carl was telling her of Adam's tenacity when he went to a desert at a certain time of year and waited days until the lighting was perfect to take the picture of the shadows on the sand dunes that he wanted. She wondered how long he waited to get the perfect picture of Yosemite. Days? Months? Years?

"Alice!" Carl said loudly. "Did you hear?"

"Uh, I'm sorry. I was thinking," she said quietly.

"I know this is difficult, Alice," Imar said.

"It is, Imar. You have no idea how difficult." Alice took a deep breath. "I don't think I have a choice."

"You always have a choice, baby," Carl said.

"I do, but it isn't much of a choice here, Carl," Alice replied. "Either I wait and see what happens, allowing the virus to continue to spread with unknown results or I take the enzyme to possibly stop its progression."

"We know the enzyme will stop the virus," Norman said.

"No, Norman, you don't," Alice said tersely. "You think it will, but you don't know."

"No, we don't," Imar replied. "Not conclusively on humans. We could conduct some more tests on animals and others in the test group, but that could take weeks or months. The virus is spreading and mutating, and waiting increases the chance of more undesirable changes. We can't keep up with it."

"So throw all common sense and procedure out the door to try a new drug?" Alice asked.

"It is not a drug or a chemical, honey," Carl said. "It is a natural element that we believe will stop the spread of the virus."

"What if it causes something else to go wrong? What if the virus feeds on it and grows even faster?" Alice asked. "Then what?"

"This enzyme is not found in this virus. It coats the virus and causes it to stop spreading, at least it has done so with all of our test subjects thus far." Norman said.

"Rats and ferrets," Alice said.

"Alice, we can't kill it because we don't know exactly how much of your brain is infected or which areas," Imar said. "We know you have cells infected by a single virus and not clusters. What we believe is that using the enzyme is about the only thing we can do to buy some more time."

"Carl," Alice said as she looked to him. He could see the question in her eyes. She needed advice, encouragement, direction.

"Honey, it's spreading," Carl said. "The virulence is significant. This is the hardest decision you will ever make, but no one can make it for you."

"What would you do, Carl?" she asked.

Carl paused as he looked at Imar and Norman, then back to Alice. "These are pretty smart scientists. We need more time. I would take the enzyme."

"What if it mutates again and spreads faster? Then what?"

"We'll cross that bridge when we get to it," Carl said.

"Carl, we're crossing a lot of bridges but not making a whole lot of progress here," Alice replied. She looked at each man and took a deep breath. "This is the only option I have, and we know it," Alice said.

"It is," Imar said dejectedly.

April 17

THE COMPOUND WAS taking shape. Owen was pleased with the design of the house and the workshops. The second and third phases were coming together nicely. He felt safe on his grounds, even though there were scores of workers milling about building walls, fences, installing concrete and block pads, raised gardens, irrigation and much more. The finishing touches were being made to the last addition to the house. It was almost complete. Four months and ten million rupees, it was well worth it.

Even with all the amenities of luxury, Owen was bored. He could only take so much television, movies, bowling and gardening. He wanted to do something, but he didn't know what. It was nice out and he wanted to get away from the constant noise of construction.

"Fadid!" Owen yelled.

Faiz appeared from the kitchen. "Yes, Meester Wallace," he said with a smile and slight bow. He had heard his name pronounced wrong so many times that he finally gave up trying to correct Owen.

"I need to get out of here for awhile. Where can we go?"

"There are many places, Meester Wallace. We can go to the market …"

"No," Owen interrupted. "Some place new, different. I want to see what this place looks like besides snow, rain, cold and blah. It's nice outside."

"It is, Meester Wallace. I suggest we go to the Paro Tshechu Festival in Paro."

"Festival?"

"Yes. We have many festivals here, Meester Wallace. The word Tshechu means 'tenth day' in your language. Many of our temples in this area have festivals to celebrate great cultural and spiritual events on the tenth day of the lunar calendar in lunar month. Today is the second day of the Paro Festival. They will have 'The Dance of the Three Kinds of Gings[27] With Drums' event."

"The what?"

"It is a spiritual dance of religion to bring good luck to all beings and to wish them happiness. This dance brings blessings to all who watch it. There will be costumes, music, dancing, food; it will be a good time for you, sir."

"Gings with drums, huh?"

"Yes, and it is in Paro."

"Paro? Where is that?"

Faiz looked puzzled. "It is the city with the airport where you landed. It is only twenty-five kilometers away."

"Oh, yeah. Right. Paro."

"I tink you will like this, Meester Wallace."

"I agree. Let's go." Owen grabbed a coat and hat from the hall tree.

"Do you need anyting more, sir?" Faiz asked.

"Just money, and I have it right here," Owen said as he smiled and patted his pants pocket.

—*—

The courtroom was full. The gallery had people standing along the back wall. The judge allowed media into the room because of the high profile of the case. A noted, reputable director of a large government agency was about to be sentenced for insider trading, fraud, and conspiracy to commit murder. Nothing like this had been tried in Washington, not at this level. The judge wanted to set an example of dealing with the corruption in Washington politics. He was itching to get this in the news.

Don Sinclair sat at the table with his defense attorney; an older woman who looked like she had been working on the farm. Her hair was white and

---

27 Gings—beings that are emanations of Guru Rinpoche, also known as the "Precious Guru" or second Buddha.

disheveled. She wore a wool jacket that looked like it had not been to the cleaners in years. She sat behind a pile of papers to the right of the defendant.

The District Attorney sat to the far right of the long, wooden table. He was a young man dressed sharply in a suit and tie. His hair was immaculate. He was a stark contrast to the older defense attorney.

"All rise." Everyone in the courtroom stood as the judge entered from the side door wearing a long, black robe. He was a balding man in his early sixties with a stern look. "The honorable Howard Wilcox presiding." The judge sat in his chair behind the elevated platform, allowing him to look down on everyone in the room. "Please be seated."

The judge began. "I want to begin by making it very clear that I expect complete silence in this courtroom during these proceedings. Any outburst of any kind will be dealt with harshly and the violator will be held in contempt." The judge flipped through a couple of pages lying on his desk. "This is case number 32-28-24776; the United States of America versus Donald Andrew Sinclair. Are the parties present?"

The DA and the defendant's attorney stood. The DA spoke. "We are, your honor."

The judge looked down and read quietly. He never looked up. "Be seated." The two attorneys sat down. "Bailiff, bring 'em in."

The bailiff opened the door to a hall behind the bench. Twelve people filed in, one at a time, and took their seats across the room from the defendant's table. The judge leaned back in his chair and watched the jurors take their seats. "Do we have a verdict?" he asked.

"We do, your honor," the woman in the first seat said. The bailiff walked over to the woman who handed him a piece of paper. He took it to the judge who opened it and read it. The judge handed it back to the bailiff who returned it to the woman.

The judge looked at Don Sinclair. "Will the defendant stand?" He looked back to the jury foreman. "Please read the verdict," the judge ordered.

The woman opened the paper. "On the charge of conspiracy to commit securities fraud we, the jury, find the defendant guilty." Don hung his head as he listened to the guilty verdict. His wife in the gallery started to weep quietly. The foreman continued. "On the charge of wire fraud we, the jury, find the defendant guilty. On the charge of conspiracy to commit murder." The foreman paused. "We, the jury, find the defendant guilty."

Don turned and looked to his wife. Tears filled his eyes as he watched her weep.

"On the charge of money laundering we, the jury, find the defendant guilty."

The foreman continued to read the charges as the audience watched. The judge made notes as the foreman read the charges. When she was finished, the judge took a deep breath.

"So say all of you?" he asked. Each person on the jury stood, one at a time and said, "Yes."

The judge looked at the papers on his desk. "Donald Sinclair. The jury has unanimously found you guilty of all charges say one. You have requested immediate sentencing. Are your affairs in order, sir?"

"They are, your honor," Don said quietly.

"Good, because I believe you need to be put away, sir," the judge said as he leaned forward and continued, "... for a long time. What you have done to so many people is unconscionable. Your greed has caused despair and misery for thousands, if not hundreds of thousands of people. Your greed and desire for wealth caused you to push a product through the review process meant to protect people. Personally, I'd like to see you charged with more counts because of the pain and suffering your actions have caused in so many lives."

Don's attorney smiled. "Your honor. Please keep in mind this is Mr. Sinclair's first offense ..."

"I don't care, counselor," the judge snapped. "And don't tell me what to do." The lady lowered her head as the judge continued. "Your defendant is a killer, a scourge on this society, truly this country. The results of his actions are devastating to so many people. However, I am only able to administer punishment based on the charges before me, and the guidelines established by this jurisdiction.

"Mr. Sinclair. You have been found guilty of all charges, say one, by a jury of your peers. Before I announce sentencing, have you anything to say?"

Don looked at the jury, then turned and saw the people in the court-room. Some of his staff were standing toward the back. He thought of bet-ter times when they worked together issuing releases for products that saved people. He saw Carl Kruger standing tall in the gallery with Imar and Norman standing on each side. He thought of the work they did together on 19Q and the lives they saved. He saw his wife, sitting next to her brother with his arm around her. "I'm sorry, Janice," he said to her. He looked back to the judge and lowered his head.

"In that case, I sentence you to the following; Counts one through six—conspiracy to commit securities fraud—seventy-two months incar-ceration and a fine of thirty seven million dollars. Count seven; conspiracy to commit murder—one hundred and twenty months incarceration. Count eight ..."

The judge read each count and accompanying sentence. Don couldn't hear anything. He shut down as his life was taken away from him and placed in a prison. He was losing everything. Cameras clicked in the gallery as Don hung his head.

When the judge finished, he closed the large file in front of him and handed it to the bailiff. "Mr. Sinclair. The accumulation of sentencing, if you were adding this up, is thirty-six years. I believe you should receive more, but I am limited to the sentencing guidelines imposed on this court. Sir, you will have plenty of time to consider what you have done to this society and this great nation. Bailiff, please take Mr. Sinclair into custody. This court is adjourned." The judge banged his gavel, stood and exited the courtroom as the bailiff took hold of Don's arm and escorted him out of the room.

—*—

Carl was sitting at his desk holding a letter in his hand and staring out the window when Imar entered. Imar noticed Carl was oblivious to his pres-ence and watched him curiously for a few seconds. Carl never moved. He held the letter and stared. "Carl," Imar said.

Carl jumped when he heard his name. "Imar. You startled me."

"So I see. You appeared to be deep in thought." Carl glanced at the paper in his hand, and handed it to Imar.

Mr. Kruger:

You have destroyed my father. He was a great man that took care of his kids and worked hard. Now, he can't do nothing but sit in a chair and stare. His mind is gone. You killed him.
Now it's your turn.

A Son.

"When did you get this?" Imar asked as he handed the note back to Carl.

"In the mail, yesterday."

"Carl, it's probably an empty threat."

"I don't think so, Imar. I got a couple of phone calls at the house, and this isn't the first threat."

"Calls?"

"This is the clearest threat. He says he's going to kill me. I'm concerned about my family, Imar."

"You need to tell the police."

"They won't do anything," he said as he tossed the letter onto a stack of mail. "There's too many people affected by this. Look at this," he said as he pointed to the stack. "Most of this are people upset about their loved one's failing condition asking for help." Carl turned to the monitor. "And the emails and phone calls; it's just too many, Imar. It's getting worse. And we can't get volunteers to test the enzyme."

"I know. They're afraid."

"I'm afraid, Imar," Carl replied. "The threats, Alice's condition, Don's sentencing, Owen's disappearance; Imar, it's overwhelming. My world is falling apart."

"Our world, Carl." Imar sat in the chair across from Carl. "I have been here for four months. I miss my home, Carl."

"Imar, don't leave," Carl pleaded.

Imar laughed. "No, I'm not planning to. I'm just saying you are home; I am not. You have Alice and Joey, I have no one. You are my closest friend. Carl, we cannot stop, even if our world falls apart."

"I don't plan to stop, Imar. I just don't know how to continue."

—*—

The drive to Paro was beautiful. Owen and Faiz sat in the back seat of the Mercedes and watched as the hills, mountains, river and plants zipped by. The winding road made Owen a little nauseous, but the trip was not far. In less than an hour they could see the edge of the city and the colorful decorations on the buildings and trees. People adorned in colorful dresses and jackets strolled along the walkways looking at the crafts and wares of the street vendors. Owen could easily spot the tourists with their slacks, sweaters, cameras and purses interspersed among the locals. They all looked alike with their fair skin and sunglasses.

Faiz pulled the car into a large lot where a few other cars were parked. A man walked up and said something in the native language. Faiz responded and handed him a wad of rupees. The man smiled and stood next to the car.

"Who is that?" Owen asked.

Faiz smiled. "He is now our guard for the car."

Owen could hear drums in the distance toward the temple on a hill. They walked across the lot to the edge of a large expanse. "Beautiful," Owen said as he saw the large temple situated on a hill surrounded by high mountains in the background and blue sky with billowy clouds. The open area around the temple had thousands of people milling about. They were dressed in colorful wraps, both men and women, with children in jackets and pants and robes. Owen noticed the men had short, dark hair; the women long, straight, black hair. Few people had gray or long hair, and almost no one had a beard. Only the very old men had beards, and they were gray and scraggly. "I feel a bit out of place here," Owen said.

"Not to worry, Meester Wallace. These people will not see you as different," Faiz said with a smile. "They will see you as a brother."

They walked past large rugs depicting religious scenes, colorful art telling a story. Owen could hear the drums getting louder the closer they got to the temple. Incense filled the air. He could see people wearing masks depicting deer with horns, buffalo, bears and other animals. He saw masks of dragons, evil and good spirits, and other colorful characters and costumes. They followed a line of people across a small bridge over a creek that flowed into the river. The bridge was made of logs tied together with ropes and was able to support the weight of the people. "Interesting," Owen mumbled. At the end of the bridge sat an older man with short, gray hair. He had weathered, wrinkled skin and was wearing a maroon knit cap, quilt coat, and a purple wrap. He was sitting on an orange blanket on the ground. In one hand he held a drum with two beads on strings. He twirled it to make a rapid drumming sound. He was ringing a bell in the other hand. He looked at Owen with the one eye he had. The other eye socket was shut with scar tissue. Owen thought he must have had an injury or infection, robbing him of his sight. How sad. Yet, the man was singing and smiling.

Owen could see people sitting on the hillsides with their families watching the events as he and Faiz continued toward the temple. They followed a procession of people up the stairs and into the square. Hundreds, if not thousands of people surrounded the square. They were almost sitting on top of each other, and more piled into the area. It was a mass of color. In the center of the square several female dancers had large dresses puffed out making them look like a top as they spun in circles to the sound of the drums. A line of men sat at the edge of the square and beat on drums of varying sizes. Lines of monks in orange wraps sat on the ground in the center of the temple square watching the activities. A long table on one side of the square held fruits, beads, and other gifts to the gods and spirits. It was a visual extravaganza.

"This is remarkable," Owen said over the beating drums.

Faiz smiled. "As I said, Meester Wallace. I tink you will like this." He motioned for Owen to continue through the crowd to a place where they could sit and watch the activities. The music and dancing continued as they each settled into their very small space. Faiz looked the crowd over and smiled. Then he noticed a small group of people directly across from him. It was Shamar and some of his men. They were watching the activities.

Owen smiled as he watched Shamar and the dancers. He wondered what blessings the dancers would bring to him.

—*—

Alice was feeling pretty good. Her memory didn't seem to be deteriorating. In fact, she felt she was improving. She had the occasional issues of forgetting where her keys were or that she left a light on. Those were expected. What scared her were the times she forgot what she was doing. She would walk into a room and stop and stare. She would have no idea why she walked into the room, or go to the refrigerator, or start to call someone and completely forget. Those times terrified her. But they seemed to be much less frequent and not nearly as intense. Even Carl commented on how well she was doing.

Maybe things were turning around and they would have time to find a cure. Maybe the inhibitor was working.

Forgotten

May 22

OWEN STEPPED OUTSIDE the tall, ornate doors and onto the wet pavement. The air after the rain shower was fresh and sweet. The stone driveway glistened in the warm sun. "What a beautiful day in paradise," he said.

"Eet is, Meester Wallace," Faiz agreed. He was sitting on the bench to the side of the entry door. "The flowers are popping their heads out and looking good, sir."

Owen looked to the daffodils and crocuses in the lower bed surrounding the fountain. They were a burst of purple and yellow. "Yes, pretty." He took a deep breath and looked toward the valley. The river meandered along the hillsides and dropped into the Thimphu Valley.

"I tink we should go for a ride in the mountains, Meester Wallace. I can show you some very interesting sights."

"That sounds like a plan."

"Good. I will have some lunch made and we will go for, what you call, a peek-neek."

Owen laughed. "Yes, a 'peek-neek' sounds like fun," he said mocking Faiz's accent. "I'll grab a sweater." Owen turned and walked into the house.

"Yes, Meester Wallace," Faiz whispered. "Eet will be most fun," he said with a smile as he turned and walked toward the kitchen to order the lunch.

—*—

The wind was blowing Carl's clothes and beard as he stood inside the decontamination chamber to the lab. It seemed like it was taking hours when, in fact, it was a few seconds. As soon as the wind stopped and the door clicked, Carl burst through into the lab. Several analysts working on their projects stopped and watched as Carl walked briskly through the main lab to vault two. Through the glass walls he could see Imar talking with Norman and another person. Imar was animated as he spoke, raising his arms and pacing.

"Imar!" Carl said as he burst through the door.

"Carl! It's progressing!"

"How can that be?"

"We don't know," Norman said.

"I thought the inhibitor stalled it."

"It did, Carl. But now it's progressing again. It has become quite virulent and is now actively killing the cells. All of the cells. Look." Imar moved to a microscope with slides on it being projected to a large screen in front of him. The slides moved on the screen as Imar moved them on the microscope. He focused in so Carl could see the images clearly. He pointed to the image on the left. "This is a sample that we took from a patient last month. Healthy neurons. Here is a sample from yesterday from the same patient." Carl could see multiple dead, colorless cells. "The virus is inside each cell, Carl." Imar zoomed in until Carl could see the virus inside the dead neuron. "It is spreading and killing the cells."

"Did this patient have the inhibitor?"

"Yes, and we have confirmed this with three other patients, Carl," Norman said.

"Three?"

"Yes. Three," Imar confirmed.

"No. There must be a mistake," Carl said. "You need to check your results again. This can't be …"

"Carl!" Imar shouted. "It's killing the cells. It's spreading."

"We've confirmed it, Carl," Norman said.

"Is it killing any other cells?" Carl asked.

"Yes," Imar said.

"Then we have to use the antidote and stop it once and for all," Carl said.

Imar motioned to the analyst in the room with them. The lady quietly left the men alone. Imar stood next to Carl. "We can't Carl. It doesn't work on these viruses."

Carl froze. He hadn't felt this terror since the 19Q project. "No, don't say that."

"Carl. We don't know how to stop it," Norman said.

Carl stared at the images for what seemed like minutes. The analysts outside watched from their stations as Carl slowly sat in a chair, never taking his eyes off the images of the viruses. "What have we done?" he said meekly. "We should have stopped it when we could." No one said a word as the news sank in. Carl remembered seeing the many patients over the years at different levels of cognitive impairment. At Golden Pond Memory Care he saw people unable to do even the slightest tasks. They were locked away with no hope and Carl couldn't bear the thought of that happening to the millions that were inoculated. He looked at Imar. "What are we going to do?"

"I don't know, Carl," Imar said. "Keep looking for an antidote, some way to stop it."

"There must be some way to break through the shell and destroy the proteins in the virus," Norman said. "We just need to know the right combination. I know we can do this."

"Even if we do, we'll have to get FDA approval," Imar said.

"Until then, how many will suffer? Millions?" Carl asked.

The severity overwhelmed each man as they sat silently stunned.

"Alice," Carl said. He looked back at the slides. "I don't think I can tell her, Imar. This is my wife." Carl started to get choked up. "I created this."

"No, Carl! You did not create this," Imar said. "*We* did." He leaned forward to see Carl's face. "I am just as much at fault, Carl. Alice is my friend. *You* are my friend. I don't want to see either of you hurt." Tears were forming in Imar's eyes. "Yet, here we are."

"She's my wife. I don't think I can tell the one I love that she is going to completely lose her memory in a very short time."

"We don't know that for sure, Carl," Norman said. "It depends on the virulence with each patient, the place of infection, propagation[28] rate, and other variables."

Carl wasn't listening. "I *have* to tell her. No one else should." Carl rose and started toward the door.

"Carl, would you like me to come with you?" Imar asked.

"No. I … uh … I think you and Norman should keep working on an antidote. We have to find a way to stop this." Carl started toward the door again when Imar grabbed his arm and stopped him. "I'm sorry, Carl," Imar said.

Carl looked at Imar with tears in his eyes. "I am too, Imar," he said, and left.

—*—

The Mercedes rounded the corner on the mountain road and continued to climb. The sun was shining as Faiz piloted the vehicle along the ridgeline above the Thimphu River. He wanted to drive the car today so he gave the driver the day off. The view was breathtaking. The road was treacherous, but Faiz knew the country well and was confident with his abilities. They crossed several of the tributaries that fed the Thimphu River from the Himilayan Peaks. They stopped at several waterfalls to take in the beauty of the rainbows reflected in the vapor of the falls. The vegetation was lush and green. The dramatic hillsides and cliffs plunged into the ravines creating multiple vantage points of view.

Faiz drove to the top of Cheri Mountain and stopped at an outcropping near the edge of one of the turnouts. He parked in the shade of a large tree. Owen stepped out and looked over the edge of the turnout down into the ravine. "Quite a drop," he said.

"Yes, Meester Wallace. Eet is a long way down to the river." Faiz glanced at his watch and then pulled out two folding chairs and a folding

---

28 Propagation—the spread of something

table and erected them under the shade of the tree overlooking the Tango Monastery far below. He placed a cooler on top of the table, two bowls and some spoons. "Lunch is served, Meester Wallace," he said as he opened the cooler and removed some tshoem, a native Bhutanese dish of beef stew and rice prepared by the house chef before they left. The steam wafted from the dish as he opened the container and spooned some into each bowl. He glanced at his watch, again.

"Expecting someone?" Owen said as he started toward the table.

Faiz chuckled, "No, Meester Wallace. My watch had some problem, I tink with the battery. I was just checking to see if it worked."

Owen sat at the table. "Does it?"

"Does it what, Meester Wallace?" Faiz asked.

"Does it work? The watch."

"Oh, yes sir. Eet does," Faiz said as he pulled up the other chair and joined his boss for lunch.

The two men sat at the little table in the middle of nowhere eating their superb lunch when a pickup soon rounded the corner unannounced. Owen watched as the truck veered toward them and slammed on the brakes, sliding to a stop not far from their table. Owen was startled as he jumped up from his chair knocking it over backwards. Faiz quickly stood and started toward the truck speaking in Dzongkha. Owen tried to understand what they were saying, but could only catch a word or two. Suddenly, one of the men stepped out of the truck. He was holding an assault rifle and aimed it at Faiz. Faiz raised his hands as the man started shouting in his native language. The driver stepped out and aimed a rifle at Owen. He motioned for Owen to put his hands up.

Owen raised his hands as directed. "What's going on, Faiz?" he yelled.

"Quiet," the driver ordered as he approached Owen.

The other man continued to point the rifle at Faiz. Faiz was terrified, looking around and glancing at Owen. The man poked Faiz with the rifle. He poked him again, in the stomach, causing Faiz to double over and fall to his knees.

"Stop it!" Owen yelled. "I have money ... rupees. I have rupees," he said.

The men looked at each other as Faiz straightened up. The man near Faiz smiled as he started to walk toward Owen and the other man. Faiz saw him moving away and bolted. He ran around the back of the truck and toward some large bushes. The driver turned and pointed his rifle at Faiz.

"No!" Owen yelled as a shot rang out. Owen could see Faiz fall forward behind the bushes. "No!" The driver slowly walked over to the large bushes. Owen watched as the driver smiled, said something, and pointed the rifle to the ground. Two more shots rang out.

"Oh, God," Owen said as he stood there with his hands up. The other man had his rifle trained on Owen and never looked away. "Yes, American. You call your God to help you now."

Owen started to shake as the reality of the situation hit him. Faiz was dead. He was alone in a foreign country. He had no one to help. "I have money," he said again.

"Quiet, American," the man said as the driver walked up.

The driver had a big smile showing his dirty, crooked teeth. He walked close to Owen and pointed his rifle at his face. Owen squinted his eyes, waiting for the shot to blow his head off. He tried to keep them open, but he couldn't. "Don't, please," he said as he started to cry.

The men laughed as the driver withdrew his rifle. "He's crying. Poor baby." He pushed Owen backwards. "Are you scared?" He pushed him again. Owen stumbled backwards and tripped over a tree branch lying on the ground. He fell back full force onto the ground and hit his head on a rock. He laid there motionless.

"You stupid!" the other man yelled in his native language. "We need him alive!" He dashed to the truck and got a canteen. He walked back to Owen and poured some water on his face. After a few seconds, Owen coughed and awoke, startled. The back of his head was bleeding. The man with the canteen pulled a bandana from his pocket and tied it around Owen's head. "In the truck," he said as he motioned to the vehicle.

Owen was dazed as he stood and wobbled slightly. The passenger grabbed his arm and helped him to the truck. Owen could see Faiz's legs sticking out from the bushes as he climbed into the back. "Hands," the man said. Owen held his hands out and stared at Faiz's legs as the man tied his hands together, then tied them to a ring in the truck bed. Owen winced as he sat on the floor of the truck bed and watched as the men walked over to the Mercedes. They looked inside to see if there was anything of value. They took a cell phone and some papers. Then they walked to the folding table and sat down. They acted like they had just been seated at a restaurant as they partook of the remainder of the tshoem. They laughed and talked as they ate the fine stew. One man held a spoonful up for Owen to see. "Thank you, American," he said as the other man laughed.

Owen periodically looked over to see Faiz's legs protruding from the bush. He scanned the roadway along the ridgeline looking for any sign of another passing vehicle. None.

The men finished their lunch in just a few minutes and walked back to the truck. The passenger tested Owen's ropes. "Good," he said with a smile.

The two men climbed into the cab of the truck and started the vehicle. Owen sat in the back and watched as the Mercedes, table, two chairs and Faiz's legs slowly vanished out of sight.

Owen lowered his head and started to cry. He realized his pants were wet.

—*—

Faiz could hear the truck drive away. He laid on the ground for a few minutes making sure that the truck was well out of sight. He slowly crawled to his knees and peered over the bushes. He watched as the truck meandered its way along the ridgeline to the top of Cheri Mountain. He smiled as it vanished over the top of the ridge.

Faiz stood up and brushed his pants off. He saw the two cartridges from the rifle lying on the ground. He stooped over and picked them up. He saw the other cartridge where the truck was parked. He picked that one up, too, and shoved all of the casings into his pants pocket. He walked over to the table and took out another bowl and spoon and placed them on the table next to the other two bowls. He dished some Tshoem into the bowl, sat down, and took a deep breath as he took in the view. He pulled out a bottle of beer, admired the label and opened it.

"Cheers, Meester Wallace, you peeg," he said as he lifted the beer and took a big gulp.

Faiz sat at the table and enjoyed his meal, alone.

May 24

JOEY WAS SOUND asleep lying against his mom on the sofa. Carl was staring at the television watching the people run across the screen, not paying attention to the plotline. Alice was snuggled against Carl trying to keep her eyes open. Carl was surprised that Alice took the news about the virus rather well. It seemed she was resigned to whatever happened. She had no control over it, and from her perspective it was her fault anyway. She had the inoculation without consulting Carl and now she was suffering the consequences. She had no one else to blame. She wouldn't do that anyway.

Still, Carl couldn't shake the feeling that he was ultimately responsible for Alice's dilemma. He wanted to fix it, but he couldn't.

"Carl?"

"Yeah, baby," he answered quietly, careful not to wake Joey.

"This feels so good, just lying against you with Joey on my lap. I feel safe."

"You *are* safe." Carl's deep voice was soothing to Alice's ears, and she smiled. "I love you more than you could ever imagine, baby."

"Ummmm." Alice squeezed Carl's arm that was lying across her chest. "I am so glad we found each other."

"Me, too."

Alice could hear Carl's heartbeat as she scooted closer and laid her head on his chest. "Do you think God knows what's happening here?"

Carl was startled by the question. "What? What do you mean?"

"You know. Does God know all of the people that this thing is affecting? All of the suffering?"

"Uh, yeah. I'm sure He knows about it."

"Then why is He letting it happen?" she asked innocently. Alice raised her head to look at Carl. "Seems like He would make it stop."

"Honey, God knows everything, but He doesn't get involved in everything. We don't know His ultimate plan, every detail. All we know is that He's saved us, that we believe He cares about us, and that His purpose is bigger than us." Carl paused. "He was willing to let His Son die for us to be saved. It's faith that gets us through anything, Alice. Faith."

"Hmmm." Alice snuggled back against his chest. "I'm scared, but I'm not scared. Know what I mean?"

"Yeah, I think I do." Carl wished he could just take her fear or pain away, and he couldn't.

"As long as I know you will never leave me, I'm not scared."

Carl hugged her. "Alice, I will never leave you. Never."

Alice sighed as she listened to Carl's heart. The television droned on and Carl thought about what the future might hold for them. He wondered how fast the disease would spread and destroy her memory. The phone rang. He reached over and saw the blocked ID. He turned the phone off. "Who was that?" Alice asked.

"No one. They must have hung up."

"Good," Alice said as she relaxed and soon fell asleep while Carl stared at the television.

—*—

The driver walked up to the small police station outside Thimphu and handed a note to the officer inside the door, then turned around and left. The policeman noticed the envelope was addressed to the Officer in Charge, so he quickly went into the next office and handed it to him. The Officer in Charge was an older man, somewhat heavy, with a thin, gray mustache. He examined the envelope that the young officer gave him. "What is this?" he asked in his native language.

"I do not know, Sagar. A man walked in, gave me the envelope, and left."

"Did you recognize him?"

"I have seen him before, but I do not recall where."

Sagar smiled. "An officer in the Royal Bhutan Police should be able to remember even the finest details of every person they meet. How old was the man?"

"Maybe twenty-six."

"Height?"

"One hundred and seventy centimeters, I guess."

"You guess?"

The young man held his hand to his nose. "Yes, one hundred and seventy."

"Weight?"

"Seventy kilograms."

Sagar was enjoying the drill. "Dialect?"

"Dialect? He didn't say anything. No dialect, Sagar."

The older man thought for a second. "Clothes?"

"Why are doing this? I have work to do, sir."

The older man stood and glared at the young man. "This is work! This is training." He stood in front of the young man. "Clothes?"

"Green shirt, short sleeve, jeans, boots, military type, and … uh … a hat. Beret type. Green."

"Unusual attire." Sagar started to open the envelope when the young man interrupted him. "I also saw a tattoo on his right forearm. It was a Tibetan circle of words."

"Tattoo? What were the words?"

"I believe it read, 'Strength—Fear—Freedom'."

Sagar stroked his mustache. "Shamar!" He quickly opened the envelope. Inside was a note.

> *"We have the American from Thimphu in our possession. His name is Owen Wallace. Send 10,000,000 rupees for his release, or he dies. We will contact you in two days."*

Sagar held the note up and looked at it. "What does it say?" the young man asked.

"It says we have a problem."

May 25

THE CHIEF OF Police of the Royal Bhutan Police force was one of the oldest men in the force. He joined the force when he was twenty, which was almost forty-five years ago. Many expected he would have left his position years ago, but he loved the power and the recognition that came with the job. No one questioned his authority. He was stocky with dark eyes, weathered skin, and gray hair. He had no facial hair because of the scar across his lower lip. He got it when he confronted a man wielding a knife many years ago. He was assigned to one of the festivals with a partner when the man, obviously mentally ill, began cursing and yelling. He ran through a crowd of people and stabbed several, including some children. The young policeman saw the commotion and, though he was only on the RBP force a very short time, ran across the square and tackled the man from behind. The man screamed, turned and cut the policeman's face deeply. The recruit quickly thrust his fingers into the man's eye sockets and blinded him. The man screamed in pain as the recruit grabbed the knife from the man's hand and thrust it into his rib cage, piercing his heart. Many people called him a hero. Others called him evil because he took a man's life without hesitation. He was known as Druk[29].

Druk sat on the bench in front of police headquarters holding the note he received from the outpost. "This is most interesting," he said in his native language as he held the note in his hand. The man sitting across from him was the Home Minister, Parash Gyloan, the head of the Ministry of Home and Cultural Affairs. The RBP reported to the Ministry, who was

---

29 Druk—Thunder Dragon—a leader in Bhutanese culture.

also in charge of immigration, law and order, civil administration and protection of cultural affairs. They were the ultimate authority and reported to the Prime Minister of Bhutan.

"And what is that, Druk?" the Minister asked. His sunglasses, white jacket and wavy, black hair made him look like an antiquities dealer from the Middle East more than the head of a major division of Bhutan's government.

"It is a note from our friend, Shamar," Druk said with a smile.

"Shamar?" Parash reached over and took the note from Druk's hand. "Do you know it is from Shamar?"

"No. I just suspect it was. The messenger had the tattoo on his arm."

"Huh. What is Shamar up to now? More drugs?"

Druk laughed. "He is always up to more drugs, Parash. This is more interesting."

Parash read the note. "Which American is he referring to?"

"The one who built the big house above Thimphu Chuu below Cheri Mountain."

"Yes. I have seen the house. Very big."

"Do you know who this Wallace man is?" Druk asked.

"No."

"I thought you would know every American who immigrated here. You are head of immigration, Parash."

"Because I am head of a department does not mean I know everything about it. Do you know who is speeding in town right now?"

"Yes. Your wife," Druk said as he laughed.

Parash looked at the note. "You should see who this American is."

"I will go to his house and look around," Druk said.

"You are a brilliant man, Druk," Parash said as he flipped the note back to the Chief. "I have a meeting to go to, otherwise, I would join you. I always wanted to see what he was building on the mountain."

"I will take pictures for you, Parash," Druk said with a laugh.

"Yes, I am sure you will."

—*—

The little black car drove up to the gated entrance to Owen's house. The guard at the gate walked up to the window. Druk flashed his credentials and the guard opened the gate. Not a word was said as Druk drove through the gate and pulled up to the front of the house.

Faiz was sitting on the porch drinking a tea. "Greetings, Chief" he said in his native language. "To what do I owe the pleasure?" Faiz stood and walked toward Druk who was approaching the porch.

"I am looking for information about an American who lives here."

"Information?" Faiz asked. "Why are you not asking for the American, to see if he is home?"

Druk realized his error and laughed. "You are a smart one ......"

"Faiz. Faiz Atal." The two shook hands.

"Faiz." Druk sat in the chair by Faiz and pulled the note from his pocket. "It appears the American who lived here has been taken; kidnapped."

Faiz looked shocked. "How terrible," he said. "I wondered where he was."

"Are you his friend?"

"Yes, well, sort of. I am his personal assistant. His name is Owen Wallace. He moved here last year from America. He hired me to help him settle in this country."

"When is the last time you saw him?"

"It was several days ago when he went for a walk."

Druk peered at Faiz. "He has been gone for several days, and you have told no one?"

"He is a very rich man, Chief. I thought he may have gone to a woman's house, you know, and stayed for a few nights." Faiz smiled. "He is American. He has done that before."

"Well, he hasn't this time," Druk said as he waved the note. "He was taken; kidnapped."

Faiz sipped his drink. "So you said, Chief. Do you know who took the man?"

"Would that matter?"

"No. I was just curious."

Druk looked around at the house, grounds and guards. "He is a rich man, yes?"

"Oh yes, very rich."

"Does he have friends, relatives in America?"

"He never spoke of them. No family, no wife or children, no one. He is alone in this world," Faiz said.

"Then, there is no one for us to give this ransom request to?"

Faiz smiled. "You could give it to me."

"You do not seem upset that he is gone. Are you not worried for his safety?"

Faiz smiled and clapped his hands. A man walked outside and greeted them. "What may I get for you?" he asked.

"Ema datsi[30] and ara[31] for me, and ... Chief?"

"For me as well, thank you."

"And for the Chief," Faiz said. The man bowed slightly and left. "You were saying, Chief?" Faiz asked.

Druk smiled. "So, you are not concerned for his safety?"

"I work for the man, Chief, but I do not have to like him. While he is gone, I am comfortable waiting for his return."

"And, what if he does not return?"

Faiz leaned forward and smiled. "That would be tragic, Chief." The waiter walked up with two glasses of ara and placed them on the small table between the men, and left. Faiz took a glass, held it up and looked at the color of the wine glistening in the sunlight. "Just tragic," he said as he took a sip and smiled.

The Chief laughed as he took a sip of wine. "If I did not know better, Faiz, I would say you are happy that he is gone."

"Oh, no, Chief. You misunderstand. I am happy waiting for his return."

"And if he does not return, what happens to all of this?" the Chief said as he motioned to the house, grounds, Mercedes, and wine. "What then?"

"If there is no family, no relatives, and no one here that is concerned about him, then, I guess, it would become the property of Thimphu since it would be 'abandoned', yes?"

30 Ema Datsi—Chili and Cheese dish.
31 Ara—Homebrewed wine.

Druk pondered the statement and smiled. "That is an interesting notion, Faiz." The waiter brought out two dishes steaming with ema datsi. Druk smelled the dish as it passed his nose. "That smells very good, Faiz."

"It is, Chief. The chef is a master of Bhutanese food. I am certain you will enjoy it."

Druk took a bite and savored the flavor. "Very nice, Faiz." He took another bite, then continued his conversation. "So, Faiz, you pose an interesting thought. If the American ..."

"Meester Wallace, Chief."

"Yes, if Meester Wallace was to not return, would this be turned over to the Thimphu government as abandoned property?"

"I think you would need to find that out, Chief. I am not in the government. I am just a personal assistant to a missing man," he said with a chuckle.

"Yes, you are." Druk continued to eat as he spoke. "If it is abandoned, who would take care of the place?"

"It would seem to me, Chief, that I have vast knowledge of the operation here, and would be happy to accommodate our wonderful government with the caretaking of such a fine facility."

The Chief laughed. "I am sure you would." The waiter came out and filled their wine glasses. "Who would pay for all of this, Faiz?"

Faiz put his chopsticks down and wiped his mouth. "Meester Wallace has a sizeable amount of money in the banks here, Chief. I would imagine the government would take control of those funds."

"I imagine they would."

Faiz sized up the Chief. He wanted to gain his protection and cooperation. He knew this was the moment. "He also has a sizeable amount of gold, Chief."

Druk almost choked on his dish as he heard the word. "Gold?"

"Yes. Gold. It is in a safe place of which I am the only person who knows where it is."

Druk glanced around to assure no one was listening. "Faiz, I am an honorable man. You can trust me with any information pertaining to this investigation," he said with a smile.

"I perceive that you are, Chief. I would need assurances that we could work together and that you could represent my interests to the Thrompon[32]."

"And what interests are those, Faiz?"

"That I would be able to take care of this place for the city, and that I would have your … complete support."

Druk pondered the offer. He took his wine glass and held it up. "Meester Faiz, I believe the Thrompon would be very happy to have an 'arrangement' with you," he said as he lifted the glass.

Faiz lifted his glass to the Chief's, and a partnership was formed.

—*—

The room was dark and musty smelling. The walls were made of wood, dirt and rock. Water seeped through the cracks in the wall creating small puddles of water on the dirty wooden floor. A small window near the ceiling let in enough light to see some cockroaches scurry along the edge of the wall looking for food. A spider was in its web high in the corner. The ceiling looked like it could have been wood with dirt overlay. Owen couldn't tell because it all blended together in the dark.

Owen sat on the mattress on the ground with his legs crossed. He stared at the two dirty buckets across the room. One was his toilet; the other his drinking water, if he wanted it. A gated door was next to the buckets. It looked like a wrought iron fence door that was used to lock him in. It was durable, and bolted to a metal frame that was mortared into the rock. Owen tried to break it loose, but there was no way. He was in a prison of some type, dug into the hillside or partially in the ground.

His clothes were dirty and ragged. He had only been gone a couple of days, but they were beginning to smell, badly. His head still hurt. His vision was a little blurry, but at least he stopped vomiting. The bandana was still on his head but the bleeding stopped long ago. His short hair was showing through the blood and scabs forming on his head. There was blood on his clothes and mattress. Owen winced as he leaned forward and grabbed his

---

32 Thrompon—Mayor.

stomach. It hurt almost constantly. "Ungh. Probably a damn parasite," he mumbled. He smacked his lips. He had no idea how bad a person's mouth could taste when they haven't brushed their teeth for days. His lack of hygiene made him very uncomfortable.

No one came to see him, except the man who emptied his toilet when it was full and brought him some rice, water and a little toilet paper each morning and night. He got nothing else. He sat in the room all day and night, staring.

A man came to the door with his evening rice, water and toilet paper and sat them on the floor as he opened the gated door. A second man stood back holding an assault rifle. Owen sat on the mattress and watched. The man brought the dirty buckets in to replace the old. "I have money," Owen blurted. The man stopped and looked at him. "I can get you a lot of money if you let me go."

The man said something in his native language, finished the bucket exchange, walked out and locked the gate. "I have money," Owen whispered. "A lot."

He listened to the sound of the men making their way through doors and away from him, leaving him in the dark, damp cell with the cockroaches and spiders. The stench from the foul buckets permeated the room. His clothes smelled: he smelled. He lowered his head and stared at the floor.

A shadow on the floor caught his eye and he looked up toward the small window. It must have been a bird fly by or something. He tried to focus, but his head still hurt. Then, he noticed a small inscription next to the window. He moved to the side and shielded his eyes from the light to try and read it.

"J-O-H-. John 3:16," he mumbled. "How stupid. What the hell is that?" he yelled. "Why write something that no one knows what it means?" Owen kicked at the floor in anger and pounded the mattress.

Spaghetti Sauce

June 13

THE STEREO WAS playing as Alice sat in front of the mirror curling her hair. Today she felt like a little girl wanting to play house. She took a shower and washed and dried her hair earlier this morning. Now, she was curling it. Normally, she would go to the salon, but today she wanted to try and fix her own hair. Joey was over at a friend's house, Carl was at work trying to save the world and she had the house all to herself. It felt good to putter around at her own pace and not have to worry about anyone getting in the way or demanding attention.

The music was soothing to her ears as she continued to apply bobby pins to the large rollers on her head. She thought she could hear a faint beeping sound, so she grabbed a crutch and hobbled to the stereo to turn it down. Suddenly, she realized she was hearing the smoke alarm.

She grabbed her other crutch and quickly hobbled around the corner. Her bathrobe was too long and she tripped over the fringe, falling to the floor and banging her head against the wall. She was dazed, but could see the cloud of black smoke in the kitchen wafting under the upper door jambs into the hallway and up against the ceiling. She struggled to get herself to her feet, leaning on her crutches. *Calm down*, she thought. *Get to the phone and get out.*

She made her way to the living room and grabbed her cell phone off the table. The smoke alarm was blaring. She was surprised that she was barely able to hear it in the back room when it was deafening here. She looked into the kitchen and saw the pan on the stove smoking furiously. There was no flame. She put the phone in her pocket and eased into the kitchen. The room was full of black smoke about one foot from the ceiling. The rest of the room was smoky, but most of the smoke stayed above the door jambs and hugged the ceiling. She hobbled around and turned the stove off. She grabbed a towel and took the burning pan of spaghetti sauce off the stove and tossed it into the sink. Then, she turned the water on. The pan popped and cracked as the cold water hit the melting pan. A small explosion threw material all over the counter. Alice grabbed the sprayer and started washing down the counter and pan until it stopped sizzling. The counter had burn marks on the Corian top and along the splash, but there were no fires.

Alice opened the door to the outside and turned the fans on to the stove and range. She hobbled around to open the windows and then to the blaring alarm. She raised her crutch and gave it a firm whack. The alarm flew off the wall and through the window shattering the glass. The alarm stopped screeching.

Alice walked into the living room and sat on the sofa. She felt the side of her head and realized she was bleeding. Her hair was a mess. She still had several curlers wrapped up in her now blood-matted hair. She pulled out one of the curlers that was hanging down, looked at it covered in blood, and cried.

—*—

"Carl, it's me," Alice said over the phone.

"What's wrong?" Carl asked. Alice seldom called him at work unless something was wrong. He was busy, extremely busy, working with Imar, Norman, and the Memory Project Team trying to find an antidote that would kill the new virus and only the virus.

"Carl. Something happened here."

"What? What happened? Are you okay?" Carl's concern came through his voice.

"I'm okay. I … I …" Alice started crying.

"Alice, it's okay. What happened?"

Alice sniffed back tears and composed herself. "I burned the spaghetti sauce," she said and started to cry again.

Carl laughed. "Sauce? You burned some sauce?"

Alice continued to sniffle as she explained. "I was fixing my hair for you and I was in the bedroom when the sauce caught on fire and filled the kitchen with smoke."

"Smoke?"

"And I heard the smoke alarm and I fell when I ..."

"You fell? Alice, Are you okay?"

The phone was silent for a few seconds, and then Alice whispered, "Carl. Can you come home? Please?"

Carl glanced over to Imar who was discussing the latest analysis of a protein attack on the virus with Norman. "Sure, I'll be right there." He hung up the phone and walked over to Imar and Norman. "I'm leaving for the day. I'll see you tomorrow," he said.

"It's only five-thirty," Norman said.

"I know. Alice had a little accident."

"Is she okay?" Imar asked.

"Yeah," Carl chuckled. "She burned the spaghetti sauce." Carl grabbed his jacket and left without giving Imar or Norman a chance to respond.

Imar looked at Norman. "Spaghetti sauce?"

—*—

Owen was lying on the mattress staring at the ceiling. A cockroach was crawling on his foot, but he didn't care. He was used to them crawling in his room. They were active at night and he managed to condition himself to ignore them. He also did more sleeping during the day when they were less active. But now, it was dark and he laid in bed watching the moonlight filter through the window onto the floor, illuminating the damp spots and few cockroaches that scurried about. The stains and pock marks on the ceiling looked like an airplane in flight.

"Hey," a man yelled as he opened the door down the hall. A light shone into the area. "Wake up, American." The pudgy man was wearing a light jacket, slacks and a straw hat and carrying a lantern and a book. He looked into the room and saw that Owen was awake, lying on the bed. "Are you alive, American?"

Owen turned to the man. "Yeah. Who are you?"

"I am your host, Shamar. You are my guest," he said with a laugh.

Owen sat up on the edge of the mattress and folded his legs under him. He was still wearing the same clothes he was captured in. His skin was developing blisters from poor hygiene. His mouth reeked, his teeth hurt. He had a hard time annunciating. His lips were cracked and sore. His hair was longer now and matted and dirty. Owen tried to wash himself with some of the drinking water, but there was so little brought each day. "Why am I here?"

"I was wondering the same thing, American." Shamar held the light so it would shine into the room. He could see Owen's eye drooping on one side. It looked like he had slept on it, but his mouth was drooping, too. A little slobber was running out the side.

"My name is … Owen … Owen …" Owen stopped.

"Wallace. Yes, I know your name," Shamar said.

"No," Owen blurted. "Not Wallace it's, uh, uh …" Suddenly, Owen was panicked. "I, uh." Owen rubbed his head trying to remember. "I was here when, uh … the car was …"

"You make no sense, American," Shamar said.

"Yes! I am American. I'm from uh, from Maryland, uh …"

"Yes. We know who you are, Meester Wallace."

"No! Not Wallace. I'm uh, uh, I'm from Maryland. I used to work in uh, a school, I, uh …" Owen grabbed his head. "God this hurts," he winced.

"Maybe you are stupid. That is why no one wants you," Shamar said.

"What? Who wants me?" Owen asked.

"No one, you fool," Shamar said.

"What?" Owen asked.

"We asked for a ransom. No one wants to pay for your release, American," Shamar said in disgust. "You are worth nothing."

"No. I have money. I have … I think. Yes, I …"

"I know you have money. Why do you think I picked you? But no one wants to give us money for you," Shamar said. "You are of no use to us." Shamar threw the book into the cell. "Like this book. No value, except for toilet paper." He turned and started to walk away.

"No! Wait! I can get someone to uh, to, uh …"

Shamar stopped and looked back. "No, it is too late for you, American." Shamr smiled, turned and left, closing the door behind him.

"Wait!" Owen yelled. "Come back!"

His voice echoed down the hall. He was alone. He realized no one was going to help him. No one was going to rescue him.

Owen lowered his head and saw the book lying on the dirty floor. It was a bible. He picked it up and flipped through the pages to John 3:16.

Vaccine

July 20

BOB STRUTHERS WAS in his mid-sixties, slim, and full of energy. He had been with CDC for twenty-two years after serving many years as a surgeon and hospital administrator. He moved into the government realm because of his passion to help the sick and protect people from illnesses. His wife died from Huntington's disease, a severe neurological disorder that eats away at its victim's memory, function and life. It was a slow, terrible death to witness. After she died, Bob poured himself into his job and career and was a true leader and example. He demonstrated this when he led the '19Q' project team a few years ago. The task was daunting, extremely frustrating and downright terrifying, but Bob maintained his professionalism and led the team to success.

Bob weighed quite a bit more back then. He knew Carl Kruger from the team and envied his well-built frame. He decided he was going to do something about his shape, so he went on a diet with a personal trainer and shed sixty pounds. Bob could accomplish anything if he set his mind to it. His favorite quote hung behind him on the wall.

"Tis far better to dare mighty things ... even though checkered with failure ... than to live in the gray twilight that knows not victory nor defeat."

Theodore Roosevelt—A Man of Action

Bob was a man of action.

Bob was the Deputy Director of Infectious Diseases for the CDC. Randall Sterling, the prior Director, suffered an aneurysm and died suddenly. Bob often wondered if Randall succumbed to the illness caused by the 19Q gene that killed so many people. Bob chalked it up to a coincidence; nothing more.

"Mr. Struthers. There's a Doctor Kruger on line two for you," came the female voice over the phone.

"Kruger? Huh. I bet it's Carl. Thanks, Sally," he said as he pushed some papers aside and hit the button. "Hello. This is Bob."

"Would you like to buy some wheat futures?" came the reply.

"Carl! Carl Kruger. I thought that might be you." Bob was very excited to hear from his old friend.

"Yeah. It's me. Been a long time, Bob."

"Too long." Bob leaned back in his chair and had a smile on his face. "Congratulations on the Nobel, Carl. That is quite some accomplishment for you and Imar."

"Yeah. Nobel." Carl's tone instantly turned downcast. "Thanks."

"You don't sound very excited about receiving it, Carl."

"Bob, I have a problem. A big problem."

"Carl, last time you had a problem we were working together trying to stop thousands of people from dying from a mutated wheat gene," Bob chuckled. "It couldn't be worse than that," he said with snicker.

"It is, Bob."

"What?" Bob leaned forward in his chair.

"It is worse, Bob. Much worse." Carl's voice was quiet, deliberate.

"I'd say you were kidding, but I know you too well, Carl. What is it?"

Carl composed his thoughts. "Bob, the virus we used to counteract the effects of Alzheimer's …"

"Yeah, the JC2 virus," Bob injected.

"Yeah, JC2. It's changed. It's now killing the neurons." Bob didn't say a word as Carl continued. "It's progressing as a normal virus would."

"So, the people who have been inoculated …"

"... are rapidly deteriorating," Carl finished. "They are worse than when they received the inoculation, Bob. Much worse."

Bob kicked into scientist mode. "What is their prognosis?"

"Grim. Severe dementia."

"Antidote?"

"We have something developed. It kills the virus, but in doing so, it kills the infected cells, too."

"No way to kill just the virus?"

"We're working on it, but we have no idea how long that could take. In the meantime the patients with the inoculation have a virus that is mutating and will eventually spread, killing their good cells."

Bob paused as Carl's words soaked in. "It doesn't sound good for the patient, Carl."

"No, it isn't. That's why I called, Bob. I need your help."

"With what? What can I do, Carl?"

"We need to get this approved and to market right away so we can kill the virus in the recipients of the inoculation before it mutates any further and spreads. We may have a window, maybe months. I need your support to push this through the channels to distribution."

"Carl, I'll do what I can, but we're going to need a lot of test data to get this through. You know that."

"We have some test results, but not enough, Bob." Carl took a breath. "Bob, Alice was inoculated."

The line was silent for several seconds that seemed like minutes. "I'm sorry to hear that, Carl," Bob said.

"She took the inoculation because she was high risk for Early Onset Alzheimer's. She wants to take the antidote as a volunteer test patient."

"Is she showing symptoms?"

"Yes. Pretty severe deterioration, Bob." Carl paused for a second. "We need your help, Bob. I need your help. I know you can get this through."

"I don't know, Carl. After Sinclair's sentencing, the FDA is up in arms about pushing anything through without multiple checks, cross checks, studies, pleadings; I mean it's almost stopped everything."

"We've got to get this going before the virus spreads."

"I'll do what I can, Carl. Is Imar working with you?"

"Yeah. Imar and Norman both."

"Norman Raynould. Huh. Sounds like the whole 19Q team back together again."

"Almost, We're just missing you … and Alice."

The phone was quiet for a few seconds. "Okay. I'll do what I can, Carl. Maybe we can approach this as a pending pandemic."

"More like an active one, Bob." The line was silent, again. Carl continued. "Thanks, Bob. You're a Godsend."

"We'll have to wait and see about that," Bob said.

Help

August 12

THE DOORBELL RANG at the Kruger residence. "I got it," Joey said as he ran to the door and opened it.

"Hi. Is this the Kruger home?" A young lady asked. She was average height with short, black hair cropped into a bobby cut. She had white plugs in each ear and a tiny, diamond stud in her right nostril. "I'm Summer," she said with a smile. Her front tooth also had a very small diamond stud in it, making her smile captivating.

"Uh, yeah. Dad?" Joey yelled behind the door.

"What's your name?" Summer asked.

"Joey."

"Hi," Carl said as he walked up to the open door. "You must be the caregiver."

"Yes. I'm Summer."

"Well, Summer. Nice to meet you," Carl said as he extended his hand. "I'm Carl. Please, come in."

Carl led Summer into the library as Joey went to the kitchen to get some sodas. "Thank you for coming over."

"Sure."

"Frances said …"

"Oh, you mean Francine?"

"Yes, Francine said she would send someone to help watch my wife while I am at work."

"Where do you work, Mr. Kruger?"

"Please, call me Carl."

"Carl."

"I work at Johns Hopkins. I'm a researcher."

"Interesting."

"Here 'ya go," Joey said as he placed the drinks on the table and sat on the sofa next to Carl.

"I need someone a few hours a day to just help with the workload," Carl continued. "My wife is suffering from dementia."

"Mom forgets a lot of stuff, too," Joey added.

"Yes, she does, son." Carl looked at his watch. "Isn't it time for you to start your homework?"

"Aw, dad."

"Go on. You need to stay on top of it, Joey. Otherwise, it'll get away from you."

"Yeah. You want to be a researcher like your dad, don't you?" Summer asked.

"No," Joey said as he turned and walked to his room, dejected.

Summer was shocked by his answer.

"He's having a hard time with this. Alice, my wife, has an aggressive disease."

"Alzheimer's?" Summer asked.

"No. It's, uh, still being researched." Carl took a drink of cola. "I need someone here to watch her while I'm working. Just to, you know, help clean, maybe some cooking, shopping, stuff like that."

"Sure. I'd love to help. Is Alice able to get around?"

"Oh, yeah. She's mobile, active, talks, but tends to get mixed up, confused. This is really hard for her."

Carl and Summer could hear the front door open and Alice's crutches bang against the door and wall as she walked in. A lady was helping her into the house. "Oh, Alice," Carl said as he and Summer walked toward them. "This is our neighbor, Maryanne, and my wife, Alice."

Summer looked at the woman, leaning on the crutches. She was smiling, but didn't say anything. Her nose was running onto her lip. "Hello," Summer said as she stood to shake hands. Alice raised her hand, but forgot she had a crutch wrapped around her wrist. The end of the crutch swung against a shelf, knocking a small figurine to the floor, breaking it.

"Oh. My crutch," Alice said as she looked at the crutch attached to her arm as though she hadn't seen it before. She ignored the pieces of the figurine on the floor.

"It's okay," Maryanne said as she and Summer started to pick up the pieces.

Alice looked at Maryanne with disbelief. "Who are you?" she asked.

Maryanne stopped picking up the broken figurine and stood. "I'm Maryanne, Alice. Your neighbor," she whispered.

"Oh," Alice said as she looked around the house, oblivious to the broken figurine and people around her. She saw the sofa across the room. "I'm tired," she said as she struggled across the room with her crutches, sat on the sofa, and sighed.

"I … I think I should go," Maryanne said.

"How was she?" Carl asked, touching Maryanne's arm to keep her from leaving. Maryanne looked at the young woman next to her. "Oh, Summer is a caregiver from Seasons In-home Care. She's going to help us watch Alice."

"Oh. Carl, she's not doing good at all. She was confused most of the time and was lost. She had no idea where she was, or who I was."

Carl looked at Alice on the sofa. "I see."

Summer could see the anguish in Carl's face. "I can certainly help, Mr. Kruger. When do you want me to start?"

Carl looked back at the young woman, smiling, eager to help. "When can you start?"

"I have the timesheets and narratives in the car. I just need to know Alice's routines, meds, and anything else you want me to handle."

"I think I'll go, Carl," Maryanne said as she started down the walkway.

"Thanks, Maryanne," Carl said as he watched her wave and walk away. "So, can you start today?" he asked Summer.

"Sure. Just tell me what days, hours, and stuff you want me to do. I'm ready, and Alice looks like she could use a little help." Summer looked at Alice sitting on the sofa holding the remote control, unable to figure out how to operate it. "See?" She walked over and sat next to Alice on the sofa. "Is that for the TV?" she asked.

Alice looked at Summer and smiled. "Yes, it is, but I keep getting the buttons mixed up. I know this changes the channel, but when I turn it on, the stereo comes on."

Summer chuckled. "Here." She took the controller and showed Alice where it read 'TV' and pushed the button. The television came on. "See?"

"Carl," Alice said as she pointed to the television.

"... were the recipients of the Nobel prize in Medicine last year," the female announcer said. A picture of Carl and Imar at the Nobel Ceremony was broadcasting on the screen as the woman continued. "They received recognition for their discovery of a cure for Alzheimer's and other related dementia diseases. The cure, 'Cerebtol,' was a tremendous success. More than three million people received the inoculation to either stop or prevent Alzheimer's Disease."

Carl stood to the side of the sofa listening to the broadcast. "That's you, isn't it Mr. Kruger?" Summer asked as she set the controller on the table.

Carl nodded his head but was silent, watching the screen as images of people in memory care facilities were packing up to leave, facilities being closed, sold, and remodeled. The announcer continued. "That was then ... people being cured of the dreaded disease. Then, four months ago in February, the product was recalled by the FDA. No reason was given." The images on the screen changed to people moving back into memory care facilities, being examined by doctors, and staring. "Today, WWL News discovered the vaccine was a live virus injected into millions of people. That virus has gone haywire, and is now ..."

Click!

Carl turned the television off and tossed the controller on the sofa. Summer looked at him with disbelief. "It's complicated, Summer," he said. She looked back to Alice, staring at the TV. "You can see why I could use your help, Summer," Carl pleaded.

Summer composed herself. "I would be happy to help, Mr. Kruger."

"Please, Summer, call me Carl."

"O.K., Carl." Summer took the controller. "Let's watch something else, Alice. O.K.?" she said as she turned the TV on and quickly changed channels to a nature show. Alice smiled as she watched the dolphins jumping in the air.

Carl walked out of the room without saying another word.

—*—

Bob Struthers wasn't used to this level of security. They had similar environmental controls and great security at the CDC, but nothing like what he encountered when he went to visit his old friends, Imar and Norman, at the JHU research center. He could see them on the other side of the glass as the wind blew his clothes around in a deafening roar. Suddenly, it stopped, and a green light came on by the handle and the computer voice invited him to exit. Bob opened the door to Imar and Norman waiting, smiling. "Bob Struthers," Norman said. He gave Bob a big hug, catching him somewhat off guard. Imar patted him on the shoulder.

"Well, this is quite a homecoming," Bob said.

"It's been too long, Bob," Imar responded.

"Yes, it has been awhile, Imar."

"I wish we could meet under better circumstances, Bob," Norman said.

"As do I. Can you tell me what's going on and where we are with this?"

"Sure," Imar said. "Let's go in here and talk privately," Imar said as he buzzed the door to the glass-enclosed room where the tissue samples were received and stored. "I have most of the reports and materials here to review." Bob noticed the multiple sample vials, dates, and names of donors as they walked across the room to the table.

Imar and Norman spent the next hour explaining the processes they went through over the past few months and where they stood. Bob knew a lot about the initial discovery of the virus and the accelerated review program by the FDA. He knew the basis of the virus, how it worked and was well aware of the inoculation program. Bob followed Imar and Carl's success through the Nobel Prize and read the numerous reports and papers about Cerebtol. What he didn't know was that Norman thought the mutated virus could be the result of the new chemical element created by Everett Maskill, a name that haunted Bob for years after the 19Q project.

"It doesn't surprise me that this virus has mutated. Not at all," Bob said. "After what we went through with 19Q and the gene, I can see how that could happen here."

"I know," Norman replied. "As with that, we have no idea why this is happening. It's likely interacting with some other element that was recently introduced or took time to break down the structure of the virus."

"But we know we can kill it," Imar said. "It's a relative of the original JC virus that our body fends off so well, but stronger. We figured out how to kill it and we can, but it also kills the host cell. The issue now, Bob, is we have to get it into the public's hands to kill the virus before it spreads while we continue to search for a remedy to kill the virus alone."

"But this only spreads through liquids, fluids, right?"

"Yes. And that is more than enough to create a pandemic. God help us if this goes airborne. This pathogen is capable of infecting the majority of the brain, destroying the neurons in short order, rendering the patient severely demented," Imar said. "If it goes airborne, it will be catastrophic."

"We know it's changed at least twice over the past four years, Bob. We can't afford to allow it to change again," Norman said.

"No, we can't," Imar agreed. "We need to have something in place now to stop it as soon as it is detected."

"Like meningitis," Bob said.

"Yes, but sooner," Imar confirmed.

Bob thought for a few seconds, looking at the screen depicting the virus and the dead brain cells spreading across the brain in shades of gray. "Okay. I think I can help get this through. I need to classify this as a medical health threat. That gives me some flexibility to work with the FDA and get this to market faster. Developing a cure is one thing. Stopping an immediate threat is another. They rely on different processes and supporting data."

"What do you need from us?" Imar asked.

"Conclusive proof that the antidote stops the virus in live subjects."

"What sample size?" Norman asked.

"At this point, I will go with minimal; it depends on the success rate. Make a one hundred percent success sample of fifty, and I can push it through for an initial distribution."

Norman and Imar looked at each other. "Fifty?" Norman asked.

"We can do that," Imar said

"Once we get the initial distribution, we'll have more evidence and be able to open the valve some more," Bob said. "It will take time, but each step will result in additional distributions."

"So we start with the evidence on the first fifty," Imar said. "The only way to obtain conclusive evidence is through a biopsy."

"Well, Imar," Bob replied. "I suggest you start calling patients and fire up the drill."

August 28

"I WOULD LIKE you to do it, Carl," Alice said as she watched the nurse place the syringe and vial on the tray next to her chair. Carl sat next to Alice holding her hand. His hand was sweating. "I want this to be between the two of us; no one else," she said.

The Memory Project team worked tirelessly for more than a month to find an antidote for the virus. They succeeded and with support from the Centers for Disease Control for a limited test, Alice was able and anxious to kill this thing inside of her. She could tell her memory was deteriorating, getting worse every day. She knew the virus was spreading and killing her neurons. She wanted it stopped now and suffer whatever results may come. She was prepared for the consequences. Carl was prepared. Joey was oblivious.

Carl motioned for the nurse to leave the room. He watched her leave and close the door behind her. "Are you sure you want this done?"

"I am, Carl. I have thought about this, prayed over it, and I really have no options. Either stop it now and control the damage that's already done, or wait and see." She looked at Carl. Her eyes sparkled. "I am not a patient person, Carl. You know that," she said with a smile. "I can't wait any longer."

Carl took a deep breath and opened the alcohol wipe. He cleaned an area on Alice's arm. He tied a rubber strap around her upper arm and then looked for the vein. He inserted the needle into the vial and extracted a small amount of liquid. He placed the needle on the vein and stopped.

"It's O.K., Carl," Alice said. He slowly inserted the needle into the vein and pulled back on the plunger to the syringe. Some blood filled the plunger and mixed with the clear liquid. Carl looked at Alice who was staring at him. She wasn't watching the injection. "You are one handsome man, Carl Kruger," she said with a big smile.

Carl looked back at the needle, took a cleansing breath, and pushed the plunger in.

Loss

September 3

IT SEEMED LIKE a lifetime ago, but, it was only a couple of years ago when Melanie Grimes was closing her memory care facility in California. Golden Pond Memory Care was converted into senior housing and Melanie retired. Memory care facilities all over the country were closing down and converting to new operations. The success of Lintrovil and Cerebtol made Alzheimer's Disease and severe dementia a thing of the past, say for a few vascular related cases. Melanie was enjoying her retirement. She moved to Pennsylvania to be close to her sisters and family. She enjoyed volunteering at the hospitals and rehabilitation centers in the area and just exploring the east coast. At sixty-two, she was a young senior herself, full of energy and ideas. She sat on the Board of Directors for the Senior Services Association of Allentown injecting her youth and vision into the association. Seniors were her life. She loved working with them, helping them in all aspects of their daily lives and listening to their fabulous stories of years gone by.

Now, she was asked to open a new memory care facility in the area. Her credentials spoke for themselves. The facility was nothing like Golden Pond. It was much larger being a six hundred room facility that was rapidly filling up with reservations. It was strategically located between New York and Baltimore to service a large area of the east coast. It was an upscale, secured facility and designed to house advanced dementia patients only.

They were victims of Alzheimer's, Cerebtol and Lintrovil. The world had changed, again. Welcome to Horizons Terrace.

Melanie pulled up to the circular driveway in front of the massive building. She watched as moving vans parked along the outside entrances, unloaded their cargo of beds, dressers, televisions and chairs. Each room was fully furnished to a standardized design and space in the facility. The residents would only bring their personal effects such as clothes, toiletries, books and pictures. Even the bedding was provided by the facility.

Melanie walked into the main entrance and watched as people scurried about. A young man pushed a large cart loaded with cleaning supplies down the hall. Two ladies were at the reception station learning how to operate the patient monitoring system. Another person was testing the room emergency lights.

"May I help you?" a middle-aged woman holding a clipboard said as she approached Melanie. She was wearing jeans and a flannel shirt half tucked in. Her hair was stuffed under a baseball cap with a red "C" on it for the Chicago Cubs. She had a nice smile and was pleasant.

"Uh, yes. I believe so. My name is Melanie Grimes. I'm the …"

"Oh! Ms. Grimes! Yes. Our administrator," the lady was quite apologetic. She checked her clothes. "I apologize for my dress. We …"

"Stop," Melanie said as she raised her hand. "No apologies needed. You guys are working your tails off from what I can see."

The lady chuckled. "Yes, we are, ma'am." She extended her hand, brought it back and wiped it on her shirt, and extended it again. "I'm Lily. The Operations Supervisor for wing two."

Melanie shook her hand. "Nice to meet you, Lily."

"Would you like me to show you around?"

"After I find my office, if that's okay?"

"Yes, ma'am. Right this way."

Lily led Melanie down the hall, past several rooms where people were installing beds, cleaning sinks, testing circuits, talking all the way. "We're planning on a full house by the end of the quarter," she said. "We already have reservations for two hundred and sixteen residents."

"All AD?"

"Yes, severe dementia, wanderers. They all need secured housing. This facility will be secured housing only, from what I hear. We have wings for elderly, middle age, and children."

Melanie stopped. "Children?"

"Yes, ma'am. Children. They received the inoculation for preventative reasons, ma'am. Some contracted it through normal transmission. It's tragic."

"Yes, it is," Melanie said.

"We only expect a few, but they will have their own living area, playground and other amenities." Lily continued the tour as they walked past one of the restaurants where residents will be fed. "This is the Rose Room. It is one of three restaurants," Lily said.

"Nice. I thought my office was back there," Melanie said.

"No, ma'am. You came in the side entrance. The main entrance is right through here." Lily led the way into a large room with a huge chandelier hanging in the center. "Voila!" she said as she threw up an arm as they entered the room.

A staircase wrapped around and up to the second floor on both sides of the room. A veranda circled above the two ladies. Large paintings were hung on the wall around the room. There were a few chairs and loveseats scattered along the walls with a couple of rosewood tables. "Very nice," Melanie said. "Very inviting."

"Yes, we like it, too," Lily said. "Your office is here, to the left of the entrance."

Lily opened the door and let Melanie through first. The office was large with a teak desk, credenza table, and bookshelves. A leather sofa and two chairs framed a glass top table with curved, teak legs. "Wow. This is quite nice," Melanie said.

"The investors have put a lot into this," Lily said.

"It shows," Melanie replied.

"Let me know when you are ready to see the rest of the facility," Lily said. "I'll be right outside directing traffic," she said as she left the room. Melanie walked over to the windows to see what view she had. That side of the facility was elevated above the driveway, so she could see down to the gardens and grassy area that approached the pond. Several men were installing a wrought iron fence similar to what she had at Golden Pond. She could barely see the end of the parking lot on the other side of the building. "Nice," she whispered. She walked over to the desk and sat in the high-back, leather chair. She looked at the glass top protecting the desk. She noticed the monitor, phone and gold pen and pencil set on the glass top. She turned the pen and pencil set around and looked at the engraving on it.

"Melanie Grimes—Administrator"

Melanie sat back in the chair and stared out the window at the fence surrounding the pond down the hill.

"Children," she whispered.

—*—

Summer was washing some dishes when Alice wandered into the kitchen. "Who are you?" she asked.

Summer wiped off her hands and turned to face Alice who was leaning on her crutches. "I'm Summer, Alice, your friend," she said as she helped Alice to the table.

Alice sat at the table and smiled. "You are a good friend, Summer," Alice said as she picked up the sandwich and took a bite. Some mayonnaise caught on the corner of her mouth, but she paid no attention to it. Her hair was disheveled, her blouse wrinkled. She had one shoe on.

"You did a good job dressing yourself, Alice," Summer said as she sat in the chair next to Alice.

Alice smiled. "I used to be a scientist, you know," she said as a few pieces of food spilled out of her mouth and onto her blouse when she spoke.

"Oh, here," Summer said as she handed Alice a napkin. "Don't talk with your mouth full. You could choke."

Alice chuckled. "I'm such a klutz," she said as more food spilled out.

Summer chuckled. "I think you are just trying to make me laugh," she said.

Alice opened her mouth wide. "Aaahhhh," she said and started laughing and some food spilled out onto her blouse.

Summer started laughing, too. "Alice!" she said as she helped wipe the food off.

"Aaaahhhh," Alice said as she stuck her tongue out more, causing some food to fall on the table.

"Alice! Stop that," Summer said, now getting irritated at the mess Alice was creating.

Alice was laughing now. "Aaahhh," *Cough, Choke, Wheeze, Cough.*

"Alice!!"

Alice dropped the sandwich and grabbed her throat.

COUGH—WHEEZE—CHOKE—COUGH

"Alice!!" Summer grabbed Alice and lifted her out of the chair. Alice continued to choke and cough. Summer held her up and leaned her over the back of the chair and watched as pieces of sandwich fell out of her mouth and onto the chair seat and floor. Summer pounded on Alice's back. "Spit it out," she yelled as she pounded and held her up. Alice continued to choke and cough. Suddenly, she was quiet. Her face was turning red. She looked at Summer with her eyes opened wide and a look of terror. She was gasping for air, but got none. Summer turned her around and locked her hands under Alice's rib cage. She was holding her up as she leaned Alice forward and thrust her locked hands into her abdomen.

Unghhh!

Alice continued to wheeze.

Summer thrust again, almost picking Alice up off her feet.

Unghhh!

And again.

COUGH!

A piece of sandwich flew out of Alice's mouth and onto the table. Alice took a deep breath.

"Alice! Are you O.K.?" Summer asked as she sat her back in her chair and looked into her eyes.

Alice looked at Summer's face and started to cry. Summer wiped Alice's face and tears and pushed her hair back. Alice leaned forward and laid her head on Summer's shoulder, and cried. Summer rubbed her back as she rocked her. "It's alright, Alice. I'm here. It's O.K. You'll be alright," she said as she rocked the crying woman. Alice wrapped her arms around Summer as she cried, and whispered, "I'm scared."

—*—

Bob Struthers was excited about his progress to get the vaccine into the right hands at the FDA for distribution approval. He spent tireless hours working with Carl and Imar to develop a program for large scale distribution based on precautionary application for a viral epidemic. The FDA agreed to the distribution program providing the CDC personally manage it from development to application to distribution and monitoring. It was a lot of responsibility for Bob, but he knew Carl, Imar and Norm well enough to know they wouldn't cry wolf and they wouldn't abandon him on something this serious. One thing he knew about each man was their word was their bond. He could trust them with his life and the lives of countless strangers.

He wanted to call Carl and give him the news, personally. It was a monumental day for the Memory Project Team. As they progressed, they could easily see the majority of recipients receive the vaccine within the next six months with the primary recipients being memory care facility employees and the medical professionals. Moreso, they would have a vaccine in place where, if the virus ever went airborne, God forbid, they could stop it in an instant.

"Carl!" Bob said as Carl answered the phone. "It's Bob. We got it!"

"We did?" Carl asked.

"Yes! They approved the program if the CDC manages it from start to finish."

"Great! We can do that. I know we can," Carl said. "There is no reason JHU would not want to."

"Then, let's get the records together to transfer the responsibility and reporting and let's get going on this," Bob said.

"Fantastic!" Carl said. "I'll have that together before the end of the week."

"Great! Looks like we are working together as a team, again."

"Yes it does, Bob. Yes it does."

—*—

Carl was excited about the great news from Bob. The drive home was troubling, though, as he changed the radio station several times to avoid hearing about the increasing number of people developing a resurgence of Alzheimer's Disease. He passed several billboards of ads touting a new Memory Care facility designed to **"feel like home"** or **"care for your loved ones like they took care of you."**

Finally, we can stop this, Carl thought as he pulled up to the front of his house. He was excited to share his news with Summer and Alice as he briskly walked up to the front door and opened it.

"Mr. Kruger?" came a man's voice from the side as a man stepped around the corner.

Carl didn't see anyone when he walked up to the door so the voice surprised him. He spun around expecting to see a man holding a gun at him. He knew it was going to be the man with the threats. Instead, a man was standing a few feet away holding out a piece of paper. "Who are you?" Carl said angrily as he looked at the man.

"You've been served," the man said as he handed the paper to Carl who took it. The man turned and left.

Carl opened the envelope and unfolded the paper.

Summons:

The Congress of the United States of America hereby summons Carl Alexander Kruger, a United States Citizen of the State of Maryland, to appear before the United States Congress on Monday, September 23, 2019 at 9:00 A.M. regarding the creation, issuance, production and distribution of a drug commonly known, prescribed and sold under the name **Cerebtol** …

Carl lowered the piece of paper, watched as the server drove away, turned, walked slowly into the house and closed the door.

September 23

THE TAXI PULLED up to the front of the Capitol in Washington, D.C. Carl and Imar peered out to see the crowd on the sidewalk and steps of the glorious building. There were scores of people carrying signs that read **'Cerebtol took my son'** and **'You robbed us of money and memory.'** News cameras were scattered throughout the crowd and on the upper steps of the Capitol capturing the moment for posterity.

"Doesn't look good, Imar," Carl said.

"No. Emotions like this never do."

Carl paid the driver and the two men stepped out of the car and onto the sidewalk. A young lady with a microphone ran up to Carl and thrust the microphone into his face. Another man with a camera stood in front of him, blocking his way up the steps. "Aren't you Doctor Carl Kruger?" the lady asked.

"Uh, yes. If you'll excuse me I need to ..." Carl started to walk forward and the lady jumped in front of him to stop him.

"There are thousands of people suffering from the drug you created, Doctor. Do you know what you are going to tell Congress about your involvement with the coverup when the drug ..."

"There was no coverup," Imar blurted. "At least not with us."

"Are you Doctor Spaan?" the lady asked.

"Let's go, Imar," Carl said as he grabbed Imar's jacket and helped him past the lady and cameraman.

"Doctor Kruger, some people believe you are culpable for ..."

Carl spun around and faced the young woman and cameraman. "Leave me alone," he said as he towered over the man and woman. He turned and walked briskly to catch up to Imar, who was struggling to make his way through the crowd as newscasters and cameras mobbed him. Someone in the crowd yelled, "You scum!" Carl looked around but couldn't see who it was. "Imar," Carl said as he caught up to him. He walked alongside his friend, protecting him from the crowd. "Are you alright?"

Imar struggled through the crowd as people thrust microphones into his face. "Yes, I'm fine," he said as the two of them managed to get through the crowd to the top steps of the Capitol. Carl and Imar walked up to a line of security guards and showed their subpoenas. The men opened ranks and allowed them through and into the rotunda in the Capitol. A security screening system, much like those at the airports, was manned and operational. The two men stepped through the image system which scanned their entire bodies, checked for explosives and imaged objects in their pockets simultaneously in seconds.

"Observer, media or witness?" a guard asked.

"Witness," Carl said as he showed them his subpoena.

The guard scanned down a paper. "This way," he said. He led them down a hall to a large, wooden door. "In here, sir," the guard said as he opened the door.

They walked into a large room filled with people. Across the front of the room were seven senators sitting behind a long, semi-circular table. Microphones were positioned at each seat. Their names were engraved on little nameplates in front of each of them. Across from the panel was a gallery of fifty chairs behind a long table which had six chairs. Carl could see his and Imar's name on two of the name plates on the table. "Looks like our seats," Carl said as they walked in from the side door of the chamber and took their seats. The crowd of people watched them and murmured.

"I think we should have brought an attorney," Imar said.

"Looks like it," Carl agreed.

News cameras and reporters were standing across the back of the room. Cameramen were adjusting their lenses and checking the focus and sound. Reporters were jotting notes on tablets. One person was sitting off to the side sketching people in the room. Carl noticed a woman in the back with a black dress, very professional looking. He didn't know it, but it was

Alicia Gold, the president of Antrole Pharmaceuticals. The one person Carl recognized was Don Sinclair.

Don was sitting between two officers. He was wearing a plain blue jacket and khaki pants. He looked at Carl but didn't say or do anything. It was as though he didn't recognize him. Carl jabbed Imar in the arm and nodded toward Don. "Over there," he whispered. "Don Sinclair."

The senators were seated across the front of the room behind the long bench. They were elevated slightly above the audience. The lady in the middle, Senator Gloria Oldham from New Jersey, was Chairman of the Investigation Committee on the FDA and approval process of Cerebtol and Lintrovil. She was an elderly lady of seventy, white hair, very distinguished looking. Carl thought she looked a lot like Margaret Giddings, the Chairman of the Johns Hopkins Board.

"Ladies and Gentlemen, if you could take your seats, please," Gloria said as she banged a gavel on the tabletop. "Quiet in the chamber, please." Everyone quieted down. "Thank you.

"We begin this hearing, September 23rd, 2019 at 10:10 A.M. This is an investigative congressional hearing on the approval process of the Food and Drug Administration in relation to two known drugs; Cerebtol and Lintrovil." Cameras started clicking in the back of the room. "This is an open hearing, meaning anyone can attend provided there is room. However, I want to be perfectly clear, this panel will not tolerate any outbursts of any type from any person in this room." She peered about the room checking to see who the likely troublemaker would be. Usually, they were easy to spot. This time, though, she couldn't tell. So many people were affected by these drugs, anyone could disrupt the proceedings. "With that said, we will begin."

The senator introduced the distinguished members of the panel. When she finished, she flipped a couple of pages in front of her and began. "Cerebtol is a drug developed by Timore Pharmaceuticals. Lintrovil is a drug developed by Antrole Pharmaceuticals. Both drugs used a common virus to initiate changes at the cellular levels in neurons to create electrostatic pathways in the brain synapse functions." The senator looked at Carl and Imar sitting at the table. "We have the two doctors slash scientists who discovered the virus and its remarkable capabilities. Doctor Carl Kruger from Johns Hopkins Alzheimer's Research Center in Maryland, and Doctor Imar Spaan, renowned microbiologist from Austria. Gentlemen." The two men nodded to the senator.

Carl and Imar introduced themselves as requested by stating their names, the correct spelling, and their current positions. They explained how they came across the virus with little Bonnie Howe, Stanley MacKenna Thorne, and Quinton Lemolo. They explained the painstaking testing they did with Leroy the rat and Albert the monkey. They shared about the ongoing tests of Phase One recipients, and the work they did with Don Sinclair from the FDA. The Senator and panel grilled the two doctors about their discovery. Two hours of testimony flew by.

"Gentlemen, this is all quite interesting," Senator Max Hambile said in a gruff voice. "But tell me this; how can a disaster like this happen if this level of research is being conducted? Did someone just drop the ball or something? Maybe hide a few facts?"

"No!" Carl said. He caught himself with the forcefulness of his answer. "Senator Hambile, we took every precaution to check and double check our findings. To be honest, sir ..."

"Honesty would be a good thing here, Doctor," Max replied.

Carl continued. "... we did everything we were supposed to in relation to research, testing, controlled environments and developing and testing hypotheses about the virus and its effects on people. We did nothing wrong."

"What about your relationship with Don Sinclair and Owen Pitke, Doctor?" another senator asked. The woman, Senator Alford, was younger, mid-forties, with heavy makeup and auburn hair. She was chubby, angry looking and spoke in an obnoxious tone. "Didn't you conspire with them to get this product to market as quickly as possible?"

"No, ma'am. Not at all," Carl said.

"We never conspired with anyone about anything, Senator," Imar said.

"Where is Mr. Pitke now, Doctor?" she asked.

"I have no idea ma'am," Carl said.

"No idea? Didn't you contact an FBI Agent by the name of Jamille Larson to track down Mr. Pitke?" she asked.

"Well, yes, but I ..."

"Didn't you discover Pitke flew to India and from there likely to a neighboring country?"

"Well ..."

"Doctor, please answer the question," she said.

"Yes, I did, but I thought ..."

"And, didn't you, in fact, meet Mr. Pitke in France when he was traveling to India on December 18th last year on his way to India too?"

"No! Senator Alford. No! I didn't meet him," Carl yelled.

The crowd started to mumble loudly. "Order, please," Senator Oldham said as she banged the gavel. "Order!" The crowd quickly quieted down. "Another outburst like that I will have this chamber cleared."

"Then, Doctor Kruger, you are saying your meeting with Owen Pitke in France was a coincidence?" Senator Hambile asked.

Carl paused for a few seconds to allow the chamber to become completely quiet. He wanted to be heard. "There was no meeting, sir. I saw Mr. Pitke boarding a flight to India as I was returning from Sweden. We happened to be in the same airport at the same terminal at the same time."

"Sure you did," Senator Alford said sarcastically.

The audience began to mumble again, so Senator Oldham took control. "We need to take a lunch break. This hearing will continue at 1:30 P.M. We are off the record." She banged her gavel and stood, indicating the current session was over.

People started to file out the back doors. Carl and Imar saw Don Sinclair rise to leave with the two guards who were escorting him. Don's hands were cuffed together. Carl walked up to Don and the guards. "Don," he said as he stopped in front of him.

Don looked at Carl with a confused look. "Do I know you?" he asked.

"It's me, Carl. Carl Kruger."

Don looked at him for a few seconds, and shook his head. "Sorry," he said as he lowered his head and started to walk away.

Carl reached out and grabbed his shirt. "Don!" he said.

The officer grabbed Carl's hand and bent it back. "Don't touch the prisoner, please."

Carl winced in pain for a second until the guard let go. "Sorry. I didn't mean anything by …"

The guards paid no attention to Carl as they each took an arm and escorted Don Sinclair out of the room. Carl turned to Imar. "He had no idea who I was," he said.

"We should go, Carl. Let's get some air and maybe a bite to eat. We have an hour or so," Imar said.

The two men made their way outside the Capitol to the top steps. Several news reporters and cameramen edged close to them, poking microphones in their faces and shouting questions. "Leave us alone, please," Carl said as they pushed their way through the growing crowd and down the steps toward the curb. The crowd rapidly grew and seemed like it was trying to swallow them.

Imar was walking next to Carl when he saw an older man, maybe in his late seventies, push his way through the crowd. The man looked angry as he pulled out a pistol and yelled at Carl, "You bastard!"

Imar instinctively stepped in front of Carl and raised his hand. "No," he yelled as the gun fired, striking Imar square in the chest. The bullet passed through Imar's chest and out his shoulder, striking the cameraman standing next to Carl in the stomach. Imar stumbled and grabbed his chest as he started to fall down the steps. Carl, who was shocked by the loud explosion, grabbed for Imar as he fell. The cameraman behind Imar fell backwards, dropping the camera on the granite steps. People around them screamed and started to scatter and fall to the ground. Two men next to the shooter grabbed the shooter's arm and fell sideways down the steps. Imar fell in the same direction. A second shot was fired as the group tumbled down the steps. The bullet passed through Imar's neck and out the back, ricocheting off the granite steps and into a lady's leg. Carl had a hold of Imar's jacket as he fell forward, spinning him around to face Carl. Imar was lying downhill on the steps, causing the blood to squirt out of his neck full force. Imar opened his eyes wide to see Carl. Blood sprayed on Carl and several people nearby. The crowd fanned out like a drop of water in a pond, waves reaching to the shores. Person after person dove to the ground, women screamed, men were yelling, and people everywhere were running away terrified. It was complete chaos.

The two men disarmed the shooter and held him down as officers scurried down the multiple steps tripping over people as they were trying to make their way through the crowd. Carl kneeled down, looking around at the chaos as he pulled Imar closer to him. The blood was squirting out of Imar's neck. Carl placed his hand over the artery and squeezed. "No!" he yelled as blood squirted through his fingers. Imar was bleeding profusely from both wounds. Blood was starting to come out of his mouth and run up his face as he looked at Carl. His wild hair had splatters of blood. "Imar!" Carl yelled. "Imar!"

Carl had blood all over his hands, arms, pants and shirt. Imar looked at Carl and tried to say something. His lips moved and faint whispers came out. Carl yelled to the crowd, "Someone get help! Please!"

Imar raised his hand slightly and pulled on Carl's shirt. Carl looked at his friend, blood oozing through his fingers as he held the artery. Someone placed a coat under Imar's head and lifted it gently. "Imar," Carl said. "You'll be O.K. You'll ..."

Imar shook his head and tried to speak. Carl leaned closer to hear him. "Carl, my friend," Imar whispered between coughs of blood. "It's O.K."

"No!" Carl yelled.

"Yes." Imar coughed. "I have lived ..." He winced in pain, and continued. "... a good life. I love the Lord." Blood oozed from his mouth as he gurgled.

"No, Imar." Carl looked around for someone to help. "Help us!" he yelled. He looked back to see the blood covering Imar's shirt. He placed his other hand over the hole to stop the bleeding. Imar winced.

"Carl. You are a good man," he said as he winced again.

"Imar, help is coming. Just stay with me ..."

Imar shook his head no. "It's alright," he said as he closed his eyes with a sigh. In a few seconds, the blood stopped pumping through Carl's fingers, and he could feel Imar's pulse stop.

"No! No!" Carl cried. "Imar!" Carl closed his bloody fists and pounded his legs in anger. "Why? Why did you do this?" he yelled as he looked up to see the shooter lying on the ground with two men on top of him and the police approaching. Tears streamed down Carl's face as he looked at the man lying on the ground. An officer just arrived and started to

reach for the man to handcuff him when Carl screamed and lunged forward, causing Imar's head to fall away onto the steps. "Arrrgghhh!!!"
Several people tried to hold Carl back, but his muscular build was made for
breaking tackles. He pulled four men with him as he made his way several
feet to the shooter who was terrified as he watched Carl quickly move
toward him on his bloody hands and knees. Carl grabbed and clawed for
him, smearing blood on the man's shoes and pant legs. The shooter scooted
backwards as fast as he could, kicking his legs to keep Carl away, but the
officer and one man held his hands and kept him down. In an instant, Carl
grabbed one leg by the ankle in an iron grip. "Why?" he yelled as the rage
continued to grow. Four men were trying to pull Carl away. Two were trying to pull the shooter away from Carl, but Carl's iron grip kept them
together. He continued to squeeze harder and harder with his blood
covered hand. The shooter yelled in pain as Carl's grip tightened. Suddenly,
his leg snapped. The shooter screamed in agony as Carl broke his ankle
with his bare, bloody hand.

The officer yanked on the man as Carl let go of the ankle. The man
screamed in pain as he was handcuffed and held down. Carl laid on the
steps, facedown, crying. The men on top of Carl slowly backed away, realizing Carl was no longer a threat. Carl turned back and crawled to Imar who
laid in pools of blood that was gently flowing down the steps. He carefully
picked up Imar's bloody head and placed it on his lap. Imar's eyes were
nearly closed. Carl softly stroked his wild hair and cried. "I'm sorry, Imar,"
he said as he cried. "I'm so sorry."

The crowd of people slowly moved in to see the bloody spectacle. The
news cameras rolled.

Checking In

October 12

IT WAS ALMOST three weeks ago when Imar was killed. The day and the moment were fresh in Carl's mind every second of every day. He was getting better, but the reality of the situation was overwhelming. He nearly broke down at Imar's funeral. It felt like everything was spinning out of control and now, the one close friend he had to lean on, was gone.

Alice was getting worse each day. The antidote stopped the virus, but the natural deterioration of her memory left her with less and less recall. Her short-term memory was all but gone. If she met someone new, she would ask them repeatedly who they were. Last week Bob Struthers had to introduce himself to Alice three times in the hour he met with Carl at the house. Finally, he related to her the time long ago when they worked on a project together at the CDC. She was able to recognize who he was and seemed more comfortable with him but only for a few moments. It was extremely unsettling for Bob considering he was the man who hired Alice at the CDC years ago and worked closely with her. In fact, he was the main reason she and Carl were married. If it wasn't for Bob suggesting they go to a dinner and movie one night years ago, Carl wouldn't have needed to carry Alice to the car when her crutch broke. She wrapped her arms around his neck and didn't let go.

Joey was struggling with his mom's condition, too. It didn't make sense to the young boy why his mom would forget almost everything. He was always watching out for her, turning things off that she turned on and on that she turned off. He was embarrassed when she put her clothes on inside out, or couldn't button them properly. He stopped inviting his friends over because he was tired of trying to explain something he knew nothing about. No, instead he would go to their house, or just stay out later than he should. No one cared anyway. Carl was gone most of the time and Alice sometimes didn't know Joey existed. He was glad when Carl sent him to live with his uncle for the winter semester of school. It was in the same city, just a different school district. Fortunately, Uncle Fred was amenable to shuttling Joey across town to stay in the same school.

Carl hired caregivers to provide twelve hours of care each day, except Sundays. That was when Carl would have Joey over for the day and they would watch after Alice. She was easy to manage once she was up, dressed and toileted. They could take her to a variety of places in her wheelchair. Her arms became too weak to use the crutches for any length of time or distance. When they went out, the chair was the better choice. Around the house, though, Alice was able to use the crutches and grab-bars strategically placed throughout the rooms and walkways.

If Alice started to go sideways, as Carl would call it, she was easily redirected to new things. One time Alice was infuriated that Carl wouldn't let her drive. She didn't understand that she hadn't driven for months now and that she would never be able to drive again. She yelled and fussed with Carl and Joey, who was terrified to see this side of his mom. Carl calmly pulled her purse off the seat and pulled out some lip balm. Alice watched as he spread the balm on his lips and smiled. "I'm painting on a smile," he said. Carl handed the balm to Alice when she reached for it and watched as she painted the balm on her lips and smiled. "You look great, dear," he said.

"Why thank you, sir," she replied.

Carl wondered if she had forgotten his name.

The caregivers provided Carl the freedom to continue his research and lead the Memory Project Team. He knew Alice would be safe, that Joey would be busy and that he would be free to work at finding a way to stop Alzheimer's Disease once and for all. The virus gave them lots of clues for a cure. They were experimenting with tau proteins, enzymes and electrical stimuli trying to find anything that would replicate the characteristics and effects of the virus on neurons. But without Imar, he felt severely disabled.

"Mr. Struthers is here," Carl's secretary said as she peered into his office. Bob walked past her as Carl replied, "Good." Bob turned and closed the door as the lady went back to her desk.

"Carl. Good to see you."

"And you, Bob." Carl looked at his friend from the CDC. "You didn't drive here to tell me how good I look."

Bob laughed. "No, I didn't."

"What's up?"

"We are getting reports that the vaccine is working. The people infected with JC2 are experiencing memory loss and degradation, but the virus is not spreading beyond a normal, water borne pathogen."

Carl looked at Bob with no emotion. "So, they are suffering as we expected."

Bob walked over and looked out the window. "Carl, it's a new world out there."

"Bob, you don't have to remind me."

"I'm not trying to make you feel bad, Carl. You and Imar did exactly what you were trained to do. Pitke and Sinclair turned it into something devastating. But we are bringing this back to normalcy."

"Normalcy," Bob laughed. "Normalcy. Normal amount of demented people inhabiting the earth." Bob tossed a pencil onto the desk. "Just a few million here and there."

"We are close to a cure, Carl. You know that. I know it. It's just a matter of time."

"Time."

"Your work will not go unnoticed."

Carl stood and walked over to Bob. "The last thing I want right now, Bob, is recognition. I would much rather have my wife and my friend back."

Bob was quiet for a few seconds as he thought. "Imar was a good man, Carl."

"Yes, he was. He didn't deserve that," Carl said as he started toward the door.

Bob continued to look out the window. "No, he didn't."

"I did," Carl said as he opened the door.

Bob was startled to hear that remark, and turned to see Carl at the door. "Where are you going?"

"Home. I'm moving Alice to a ......" He couldn't finish. He turned and walked out, leaving Bob Struthers alone in the office.

—*—

The van pulled up to the front of the large, four story hotel-like building. Carl thought it looked remarkably like an Embassy Suites hotel. A large sign made of stucco with brick trim and brass lettering greeted the newcomers.

Welcome to Horizons Terrace.
Home is where the heart is

Carl stepped out of the van and walked around to the other side as the electric sliding door opened. Summer stepped out and pulled a wheelchair out with her. She opened it up as Alice sat inside the van and watched. Carl helped put the foot rests on the chair and the seat pad. "There you go," he said as he slapped the pad.

Alice stared at the chair. Summer reached in and helped Alice transition into the chair. She placed her feet on the foot rests and adjusted her jacket. "I'll park the van and bring her clothes, Mr. Kruger," she said as she walked to the back of the van.

Carl pushed Alice along the flowers bordering the circular driveway. He could see the pond down below the rise with a large walkway around it. People were walking, holding hands, talking. Some were in wheelchairs being pushed. Others were using walkers. Everyone was with someone. No one was alone near the pond.

The entrance was grand. Two large, glass double doors were trimmed in brass. Inside was the main lobby with a reception desk to the right. It was designed to look and feel like a hotel. The staircases lining the lobby were trimmed in white railings. The carpet was plush. The chandelier was sparkling and reflecting prisms throughout the lobby. A wooden table with a vase and several tall, colorful plants sat directly under the chandelier. "Oh, so pretty," Alice said with a smile as Carl thought it looked a lot like the lobby at the Grand Hotel in Stockholm.

Carl pushed Alice into the lobby and looked around. Several people were either sitting in Early American chairs or walking by. There was no odor of urine or feces here. Staff was on top of every accident in seconds.

"Welcome to Horizons Terrace. May I help you?" the woman asked. Carl turned to see Melanie Grimes standing with another woman. She looked at him. "Do I know you?"

"No, I don't think so," Carl said.

"I'm sure I've seen you somewhere before."

"No. I was in the news a time or two, so maybe there," Carl said. He thought she may have recognized him from the shooting, or the Cerebtol disaster, or the Nobel Prize. He didn't want any recognition. He just wanted to be left alone and check his wife into her new home.

"Well, welcome to our new facility," Melanie said as she swept her arm in front of her. "I'm Melanie Grimes, the administrator, and this is one of my managers, Gwen Furth."

Carl recognized the name, but ignored her. He didn't want the connection. He was afraid she would recognize him when he said their names. "Ladies," he said as he shook hands.

Gwen leaned down to greet Alice. "Hello, Alice," she said as she placed her hand on Alice's arm.

"Hello," Alice said with a smile. "Do I know you?"

"Yes, I'm Gwen. We met last week when I came to your house to do the assessment, remember?"

Alice smiled, but couldn't remember. "I don't think I do."

"That's O.K., Alice. We're going to get you checked in and into your new room." Gwen motioned to the side door. "We can go right to her room if you like. We've already done the paperwork."

Summer walked through the door rolling Alice's bags. "Here's her bags now," Carl said.

The group followed Gwen through the secure doors into wing four.

—*—

"You've been very helpful," Carl said as he finished writing Alice's initials and room number on her blouse and hung it in the closet.

"You are welcome," Gwen said. "We love having new guests." She walked over to Alice and kneeled down. "If there's anything you need, Alice, you can just pull any of these chords or ring this buzzer. We'll be right here to help." Alice looked at the things Gwen pointed to and smiled. Gwen left, leaving Carl and Summer in the room with Alice.

"Well, I guess this is it for now, baby," Carl said as walked over to Alice sitting in the wheelchair. "We need to go back to the house and ..."

Alice reached over and grabbed Carl's arm. Her eyes were wide. "Don't leave me," she said.

Carl looked at her, surprised by her sudden awareness of the situation. "Honey, I'm only going to be gone a very short time."

"Carl," she pleaded. Summer left the room. Alice watched her walk out. "Carl."

"Alice, I have to go, but just for a little while. I need to take care of some things." Alice continued to look at him. "Think of this as a mini-vacation. I'll be back here in no time." Carl stroked her hair and watched her close her eyes. She was always at peace when Carl touched her. He leaned over and whispered into her ear. "Baby, I love you. I will be right back."

Alice watched him stand. Gwen came back into the room and redirected Alice. "Hi, Alice. I'm going to the restaurant to play bingo. Will you come with me?"

Alice smiled. "Yes. I like bingo," she said.

Gwen nodded to Carl to go. He walked out of the room as Gwen and Alice talked. He felt so badly. When he got to the end of the hallway near the secured doors, he looked back to see Gwen rolling Alice out of her room and down the hall away from him. Alice leaned to the side of her wheelchair and looked back at Carl.

Carl waved. Alice didn't respond. She turned around and sat in her chair as Gwen took her to the restaurant. The door buzzed, and Carl left wing four.

Year Seven—2020

One Spring Day

March

IT WAS A beautiful spring day in late March. The sun was warming the countryside as Carl and Joey enjoyed their drive to Horizons Terrace to see Alice. It was now a weekly visit for the two of them. Joey was spending most of his time with friends and his aunt and uncle, while Carl continued to work at the research facility. He didn't work the long hours that he used to. His heart wasn't in it any more. After his Nobel Prize was revoked, he was resigned to work a normal, forty hour work week and spend more time with Joey when he could, particularly on weekends.

Today, they were heading to the facility to enjoy an Easter brunch with Alice and the guests. Joey had a chocolate bunny with the ears chewed off. He knew his mom liked chocolate and he hoped she would remember when she would chew the ears off the chocolate bunnies that she used to give him. Carl and Joey were always trying to find ways to get her to remember times past. Occasionally, they were successful. More often than not, they weren't.

They drove out of Baltimore and headed north toward Allentown. They passed several billboards touting the benefits of various memory care facilities. Carl stopped listening to the news long ago. Thousands of care facilities cropped up across the country. Ads on television, the newspapers, radio and billboards continually reminded him of his involvement with the new wave of increased dementia in society. There was no solace in the fact that Congress absolved him of any wrongdoing. The class action suits were just surfacing because of Pitke's and Sinclair's negligence with the development and release of Cerebtol.

The biggest news of recent weeks was the dissolution of the two companies involved with the production of the new miracle drugs. Many pharmaceutical companies suffered as a result. The confidence with their research waned.

"Think she'll remember the chocolate?" Joey asked.

Carl smiled. "Yeah. How could anyone forget something that funny?" he said.

They rounded the corner and parked in the large lot to the side of the building. They had been coming here for months, but each time they saw the building they were impressed with how large and beautiful it was. Joey jumped out of the car, grabbed the chocolate bunny and a bag of bread crumbs. Carl got the flowers he brought and they both headed into the building to see Alice.

Alice was sitting in her wheelchair watching the residents and family members pass by. She was dressed in a pretty flowered blouse with white slacks. Her hair was done and she had a little makeup on. Carl thought how pretty she looked, even as disabled as she was. "Hi, baby," Carl said as he walked over to hand her the flowers.

Joey ran to her side and gave her a hug. "Hi, mom," he said.

Alice was a little startled, then smiled. "Hi, there," she said.

Carl knew that she didn't know who the boy was. She learned to cover well. "Joey, why don't you give your mom your present?" he said, cueing Alice about who he was.

Alice smiled. "Yes, Joey." Joey handed her the chocolate bunny. "Oh, the ears are gone," she said with a chuckle. "Where are they?"

Joey giggled. "I ate 'em," he said proudly. He was thrilled she remembered, though she really didn't. She just saw that the ears were gone.

Alice opened the bunny and took a bite of chocolate. "Umm," she said.

"Joey, why don't you head down to the pond and feed the ducks? We'll be right there," Carl said.

Joey bolted out the door and ran down to the pond with his bread crumbs. Carl bent down and gave Alice a hug. "Good to see you, baby," he said.

Alice looked at Carl. "I like your voice." She paused for a moment. "Who are you?"

Carl's heart sunk. He thought he could get used to her forgetting who he and Joey were, but it continued to hurt each time it happened. "I'm Carl. I'm your husband," he said.

"You are one handsome man, Carl," Alice said with a smile.

Carl smiled back. "Let's go down to the pond and watch your son, Joey, feed the ducks."

"O.K.," Alice replied. She was in a good mood today, and amenable to anything Carl wanted to do. Maybe it was the sunshine. Maybe it was the medicine.

Carl pushed Alice down the walkway toward the pond. They stopped at a bench under a cherry tree in full bloom along the walkway a little ways from Joey. He was throwing clumps of bread into the water and trying to get the ducks to eat out of his hand. "He's cute," Alice said.

"Yes, he is," Carl replied. Carl parked Alice's chair at the end of the bench and sat next to her, facing her. "I sure miss you, baby," he said as he held her hand.

Alice looked at Carl's hand and then into his eyes inquisitively. Suddenly, she recognized who he was. "Carl?" she asked.

Carl was shocked. "Alice? Do ... do you know who I am?"

"Carl!" Alice reached out and grabbed his arm. "Carl!" She was almost clawing her way into his arms. "Carl!" Alice started to cry as they embraced. "Where have you been?"

Carl knew that people with Alzheimer's Disease may periodically regain memories and appear normal for fleeting moments as the disease progressed. He knew, but each time it happened, it tore at his heart. He was thankful if he could be there when it occurred, and he hated when it did. "Alice, you're back. Alice." They held each other, Alice weeping gently on his shoulder. They parted for a moment as Carl looked into her eyes. He could see she recognized him. He was seeing his wife, again, and she saw him. "Oh, Alice, I miss you," he said.

Alice looked around at the strange surroundings. "Carl, I'm scared." Carl could see the fear in her eyes as she looked around in confusion. "I don't know where I am. Where am I?"

"With me under a cherry tree," he said.

"Where's Joey?"

Carl pointed to the boy by the water. "There."

Alice saw Joey by the water and smiled. "What day is it?"

"Easter."

She turned to look into Carl's eyes. "He is risen," she said with a smile.

Carl smiled widely. "He is risen, indeed."

Alice raised her face toward the sun and closed her eyes, feeling its warmth. "The sun feels good," she said.

Carl turned her head back to him. "I love you, baby, so much," he said as he reached out and cupped her face with one hand. Alice leaned into his palm and kissed it. "Carl, you are the only man I have ever loved," she said.

"Hey, look at this," Joey yelled, interrupting their moment. "I'm the Pied Piper."

They both turned to see Joey walking along the edge of the pond with a small flock of ducks and geese waddling behind him. Carl laughed. "Yes, you are. Be careful that they don't bite you," he yelled.

"They don't have teeth, dad. They can't bite."

Alice was smiling as she watched the little boy. Then, she turned to Carl. "Who is that?" she asked.

Carl realized Alice was gone, again. "Alice?"

"Yes, I'm Alice." She looked at Carl inquisitively. "Who are you?" she asked.

Carl took a deep breath to compose himself. He wanted Alice back. He fought back tears and said, "I'm Carl." He reached out and took her hand.

"Nice to meet you," Alice said with a smile. "I like your voice, Mr. Carl."

Carl continued to hold her hand as they watched the ducks and Joey. They said very little, but Alice felt safe, and Carl felt loved.

Suddenly, Carl's phone rang. "Hello?"

"Carl! Carl! We did it," the voice over the phone said excitedly.

"Did what?" Carl asked.

"We found the cure! The cure for Alzheimer's. We did it!" Carl pulled away from Alice as the voice continued. "We used a combination of tau proteins and enzymes to ..." Alice watched as Carl stood and walked to the edge of the pond as the voice continued. "... to replicate the neuron transmissions ..." Carl listened as he watched Joey play with the ducks. Then, he pulled the phone away from his ear and looked at it while the voice continued to blare. He looked up to see Alice sitting in her wheelchair under the blooming cherry tree enjoying the warmth of the sun. "... with neurogenesis and ..." Carl turned back to the pond to see a hen duck floating gracefully across the still water followed by four ducklings. "... complete reversal of synapse damage ..." Carl looked at the phone in his hand, and tossed it into the water. He watched the light of the phone slowly sink and flicker out of sight. He casually walked back to the bench under the blooming cherry tree where Alice was sitting.

"Hello, Alice," Carl said as he walked up and sat on the bench. "My name is Carl." He reached out to take her hand.

"Hello, Mr. Carl." Alice reached out and took hold of his hand. "I like your voice."

The two of them sat under the cherry tree holding hands and watched Joey play with the ducks. Alice felt like she was flirting and giggled.

It was a beautiful, spring day.

Epilogue

A CLOUD OF dust was moving through the mountains near the bor-
der of Pakistan and Afghanistan. It was nearly dark as the last remnant of a
beautiful sunset disappeared on the horizon. The small truck had one good
headlight shining down the road. The front driver side of the truck was
caved in from an old accident. Two men sat in the cab of the truck as it
bounced through the mountains. "Do you know where you are?" the young
man asked in Dari[33]. He was in his early twenties and dressed in an army
jacket and pants. A 45 caliber pistol and AK47 rifle sat on the seat between
him and the driver.

"Of course, you fool," the older man replied. He continued to stare at
the road, winding his way through the mountains.

"We must get him to the hospital tonight," the young man said. "They
have a plane ready for him."

"We will, we will. Allah will guide us, young one."

Three men sat in the back of the small truck in the open bed. They
wore dirty galabiyyas and turbans. Two were using the cab as a backrest.
The third was lying on a mat in the bed of the truck. He was the leader of
Al Tahrir. His pants revealed he had urinated on himself sometime during
the trip. His hair was matted and stringy. He stared at the feet of one man
as he drooled onto the mat.

"It is important that we get our leader to the hospital, Abi," the young
man in the cab said. "They can fix him."

---

33 Dari—Primary language of Afghanistan.

"Allah will fix him. He may choose to use doctors, but Allah will fix him. Never forget that," the older man said with a stern look.

"Where are we?" the young man asked. The driver ignored him. The young man rolled down the window and leaned out. He yelled to the men in the back, "Do you know where we are?"

One of the men turned and stood to look over the cab. It was dark and the road was becoming treacherous. "I think we take the next road left, but I don't remember."

"I think it's to the right," the driver yelled out the window.

The men started yelling about which direction to take. Suddenly the truck hit a pothole and bounced mightily, tossing the man in the back nearly on top of the cab as the headlight flickered off. The loud noise of the man hitting the top of the cab caused the driver to look up as he came to a turn. When he looked back, the road was black. He couldn't see where the truck was headed. The headlight flickered back on and the men in the cab could see the oncoming cliff. The driver slammed on the brakes, but the downhill momentum and the gravel road allowed the truck to continue in a slide over the cliff. The men's screams echoed through the canyon as the truck plummeted into the black ravine. The single headlight could be seen miles away as it floated to the bottom of the cliff and flickered off.

The End.

Afterword

THIS IS MY second venture into writing a novel. My first, '19Q', was well received. Thank you to my friends and strangers who commented openly and honestly on my effort and encouraged me to write another book. It gave me a good feel for where I needed to change both in style and detail.

This book took four months to write. When I wrote '19Q,' I planned to have a sequel that took the wheat mutation to a virus that ultimately went airborne. The problem was that 'Contagion' was released right after I finished '19Q,' so I was stuck.

I decided to explore Alzheimer's Disease; a terrible, terrible disease that cripples so many people, destroys families, and fills our hospitals, nursing homes and care facilities. We are learning more about it every day, but there is much more to learn. There is a cure out there somewhere; we just haven't found it yet. I believe when we do, we will see the same level of excitement and display that I depicted here. The Nobel Prize, news, changes in elderly housing will be but a few of the effects of such a great discovery.

I wanted to use the possibility that we, as a society, could be so excited about something so desperately needed that we could push it through the process to benefit so many people, only to have it turn into a disaster. We have checks and balances in many areas of medicine, but ultimately, we have to take risks each time something is released, not knowing the long-term effects of the product.

The work I currently do is acting as a professional fiduciary appointed by the court for people unable to care for themselves. I see the effects of AD every day. I see the families torn, the children and strangers taking advantage of an elderly person who only wants to stay in their home so they trust anyone, never expecting they would steal them blind. This is truly a scourge on our society. I, for one, will be very happy when a cure is found and this terrible disease can someday be a thing of the past.

The scene with Mac entering the lobby at a care facility and singing the Lord's Prayer was an event that I actually witnessed. Many years ago I went to a care facility and watched as a man who looked like Mac did exactly that. I was astonished to hear him sing. The entire facility stopped as he sang. This was the inspiration for that scene; a true story personally witnessed by me.

Another scene that actually happened was the yogurt spill. My son and nephew spilled some yogurt on the floor and I decided the best way to clean it up was to lick it up like a dog. That was one of the most fun moments I can remember being a dad: the three of us licking and barking like dogs, laughing all the while.

The purpose of this book was to, hopefully, entertain, but to also convey my thoughts about God's involvement with our lives. Too often I hear people ask how God could allow *that* to happen. It occurs when a child is murdered, or a home destroyed, or someone is crippled, maimed, and so on. Look at the sinking of the Titanic one hundred years ago. Men, women, and children lost at sea. How could God allow that to happen? No one knows God's thoughts. We may think we do, and we may have snippets, but who are we to question something just because it doesn't go our way? In this story, Alice suffers, millions suffer, yet that is what it took to find a cure in the very end. It also stopped an extremist cult from causing more harm in the world. Maybe they were going to be responsible for killing the next president of the United States, or an inventor of something world changing. We don't know; *and we don't need to know.* We are called to believe. Faith.

One of the most important points of this book was one sentence uttered by Alice; "He is risen." I wanted to show that a person, such as Alice, could be going through a horrific situation in their lives that they literally have no control over. She states several times that she is scared. Carl's presence reassures her and calms her down. Her faith in God pulls her through at the end, giving her peace that surpasses our understanding. Alice saw that in the midst of her dilemma by Carl telling her it was Easter, and she responded with complete peace by uttering the one sentence. How appropriate that I finished the first draft of this book on Good Friday. As a Christian, I believe faith in Jesus Christ resurrected, is the ultimate, true hope.

During the most difficult times in our lives, those of us who believe in God and His son Jesus, have a hope and faith to endure. Things may not go the way we expect, and more often than not, they don't. _The crucible of adversity is the testimony of our faith._ Alice Kruger was scared, entering into a terrifying situation yet, when she thought about Christ even for a moment, she felt safe and at peace. I believe we can each have that same level of peace ... if we but believe ...

Gerald Rainey

Appreciation

I want to thank several people for their support, encouragement, and assistance.

- Leotta Emery—This wonderful, godly woman spent hours proofing my book for grammatical correctness. Thank you so much.
- Carol Sylliaasen—A beloved friend who proofread the manuscript for content flow and story line.
- Jerry Thomas—A good friend and Ph.D. in Chemistry. He proofread my book for scientific accuracy. I learned so much about viruses and bacteria from a brilliant man of God. Thank you.
- And my lovely wife, Rikki, who believes in me and encourages me to go beyond.

Bibliography

Research was done entirely on the Internet through the following links:

http://www.cdc.gov/HAI/organisms/acinetobacter.html

http://en.wikipedia.org/wiki/Long-term_potentiation

http://www.scientificamerican.com/article.cfm?id=sewage-plants-super-bacteria

http://en.wikipedia.org/wiki/Sewage_sludge_treatment

http://en.wikipedia.org/wiki/Sewage_treatment

http://www.microtack.com/html/natural_treatment05.htm

http://en.wikipedia.org/wiki/Constructed_wetland

http://www.ncbi.nlm.nih.gov/pubmedhealth/PMH0001767/

http://en.wikipedia.org/wiki/Alzheimer's_disease

http://www.nia.nih.gov/alzheimers/publication/alzheimers-disease-genetics-fact-sheet

http://en.wikipedia.org/wiki/Amino_acid

http://www.aboutintelligence.co.uk/intelligence-memory.html

http://en.wikipedia.org/wiki/Chromosome_19_(human)

http://www.bloomberg.com/news/2011-04-04/five-new-alzheimer-s-genes-double-total-in-new-push-for-cure.html

http://www.alz.org/downloads/Facts_Figures_2011.pdf

http://en.wikipedia.org/wiki/Memory

http://www.webmd.com/alzheimers/news/20110403/new-alzheimers-genes-found

http://www.clinicaltrials.gov/ct2/show/NCT01354444?term=alzheimers+and+johns+hopkins&recr=Open&rank=3

http://trials.johnshopkins.edu/?ctg=alzheimers

http://en.wikipedia.org/wiki/Activities_of_daily_living

http://en.wikipedia.org/wiki/Mini%E2%80%93mental_state_examination

http://en.wikipedia.org/wiki/Neurogenesis

http://biology.about.com/od/Brain/p/Regeneration-Of-Brain-Cells.htm

http://www.alz.org/alzheimers_disease_steps_to_diagnosis.asp

http://en.wikipedia.org/wiki/Ligands

http://en.wikipedia.org/wiki/Neuroimaging

http://www.nmr.mgh.harvard.edu/martinos/research/technologiesPET.php

http://www.mcghealth.org/radiology/GhsuContentPage.aspx?nd=565

http://en.wikipedia.org/wiki/SPECT

http://www.alzresearch.org/trial_detail.cfm?study=12

http://www.fda.gov/NewsEvents/Newsroom/PressAnnouncements
/ucm278646.htm

http://www.neurosurgery.ufl.edu/clinical-specialties/stereotactic-
brain-biopsy.shtml

http://www.txalzresearch.org/index.php?option=com_content&view
=article&id=52&Itemid=68

http://www.msnbc.msn.com/id/32937442/ns/health-
alzheimers_disease/t/dementia-toll-climbs-million-worldwide/

http://www.alz.org/documents_custom/report_alzfactsfigures2010.
pdf

http://en.wikipedia.org/wiki/Nootropic

http://www.ahaf.org/alzheimers/about/understanding/plaques-and-
tangles.html

http://en.wikipedia.org/wiki/Insulin

http://speakingofresearch.com/2009/07/13/from-mouse-to-
monkey-to-humans-the-story-of-rituximab/

http://en.wikipedia.org/wiki/Immune_system

http://en.wikipedia.org/wiki/Leukopenia

http://en.wikipedia.org/wiki/White_blood_cell

http://www.msnbc.msn.com/id/44275043/ns/health-cold_and_flu/t/why-some-people-dont-get-flu/

http://www.wakehealth.edu/Research/WFUPC/Cynomolgus-Monkeys.htm

http://www.beyondbooks.com/lif72/2.asp

http://en.wikipedia.org/wiki/Acinetobacter

http://www.nytimes.com/2011/12/11/health/research/hemophilia-b-gene-therapy-breakthrough.html?_r=1

http://www.waterbornepathogens.org/index.php?option=com_content&view=article&id=57&Itemid=67

http://en.wikipedia.org/wiki/Polyomavirus

http://www.webmd.com/alzheimers/news/20050610/immune-system-problem-linked-to-alzheimers

http://immunedisease.com/about-pi/types-of-pi/common-variable-immunodeficiency-cvid.html

http://www.cdc.gov/vaccines/pubs/pinkbook/downloads/appendices/C/vax-storage-temps.pdf

http://en.wikipedia.org/wiki/Human_microbiome

http://www.virologyj.com/content/4/1/65

http://en.wikipedia.org/wiki/Polymerase_chain_reaction

http://www.vetmed.ucdavis.edu/error.cfm

http://www.funeducation.com/tests/KidsISIQ/Kids-IQ-Testing.aspx

http://www.rubiks.com/world/history.php

http://en.wikipedia.org/wiki/Hippocampus

http://www.fda.gov/Drugs/DevelopmentApprovalProcess/default.htm

http://en.wikipedia.org/wiki/Investigational_New_Drug

http://en.wikipedia.org/wiki/New_Drug_Application

http://en.wikipedia.org/wiki/Clinical_trials

http://www.drugs.com/fda-approval-process.html

http://en.wikipedia.org/wiki/Informed_consent

http://www.fda.gov/Drugs/DevelopmentApprovalProcess/HowDrugsareDevelopedandApproved/ApprovalApplications/InvestigationalNewDrugINDApplication/ucm176522.htm

http://en.wikipedia.org/wiki/Federal_Bureau_of_Investigation

http://www.howstuffworks.com/mri.htm

http://en.wikipedia.org/wiki/Lumbar_puncture

http://www.scandirectory.com/content/brain_scan.asp

http://www.gizmag.com/feiter-brain-image-consciusness/18931/

http://www.gizmag.com/feiter-brain-image-consciusness/18931/picture/136040/

http://en.wikipedia.org/wiki/Phenylalanine

http://en.wikipedia.org/wiki/File:Spinal_needles.jpg

http://en.wikipedia.org/wiki/Louise_M._Davies_Symphony_Hall

http://en.wikipedia.org/wiki/Nobel_Prize_in_Physiology_or_
Medicine

http://en.wikipedia.org/wiki/Paro_District

http://www.hotelchatter.com/story/2008/2/11/95429/0550/hotels/
Stockholm_Hotel_Scene%3A_Noble_Prizes_Galore_at_the_Grand_
Hotel_

http://www.nobelprize.org/nobel_prizes/medicine/laureates/1978/

http://www.nobelprize.org/nobel_prizes/medicine/laureates/2009/

http://en.wikipedia.org/wiki/Transient_ischemic_attack

http://www.hotelscombined.com/Hotel/Grand_Hotel_Stockholm.
htm

http://swiss-bank-accounts.com/e/banking/bymail.html

http://en.wikipedia.org/wiki/Thimphu#Geography_and_climate

http://onlygold.com/coins/WeBuyHalfOunceCoinsFS.asp

http://www.tablethotels.com/Taj-Tashi-Thimpu-Hotel/Thimphu-
Hotels-Bhutan/115439?gclid=CMOPgMa8va4CFRJThwodfRpCIA

http://www.aircharterservice.aero/pax/aircraft/bell_222.htm

http://en.wikipedia.org/wiki/Allahu_Akbar_(anthem)

http://en.wikipedia.org/wiki/Terrain-following_radar

http://www.achodaka.com/festivals-in-bhutan

http://www.tourism.gov.bt/what-to-do/paro-tsechu-festival.html

http://www.c-span.org/Events/Senate-Energy-Cmte-Examines-Current-and-Near-Term-Gas-Prices/10737429430-1/

http://en.wikipedia.org/wiki/United_States_congressional_hearing

About the Author

Gerald was born in Sacramento California, a fifth generation Sacramentan. He grew up in the area graduating from Elk Grove High School in 1971 and eventually California State University, Sacramento. He joined the navy in 1972 and served on and aircraft carrier: the USS Enterprise. He did tours in the Asian Pacific, Viet Nam, and the Persian Gulf.

After leaving the Navy, he met Rikki, his wife of thirty-five years. They have one son, Joshua, and one grandson, Kellen.

He testified against two gang-members who shot at him in his front yard after they burglarized a neighbor's house. The gang-members were sentenced to prison and Gerald moved his family to Oregon.

Gerald worked in the credit union field for twenty-five years, eventually leaving his last position as Chief Financial Officer for a billion dollar credit union to start a new business with his wife in 1999: Cornerstone Services, Inc. They act as professional fiduciaries, appointed by the court as conservator, guardian, trustee or personal representative to manage extremely difficult cases.

In 1988, Gerald's first book was self-published. It was a self-help guide to finances called "Free From The Love of Money." His second work, "19Q," was published in 2011. It is a science fiction thriller about genetics gone bad. The Memory Project is his third work.

Gerald is active in church performing in plays, playing drums on the worship team, and general volunteer duties. He also teaches church business administration as an adjunct professor for New Hope College in Eugene, Oregon. During his downtime, Gerald likes to explore other activities, such as hunting, fishing, road trips, travel, writing, music, drama, and just visiting with family and friends. He also likes to tinker with cars, having restored a 1963 Ford Galaxie with his daughter-in-law, Rachel.